# Laboratory Exercises to accompany

# Invitation to Oceanography

# *Laboratory Exercises to accompany*

# Invitation to Oceanography

Paul R. Pinet

Prepared by

## Karl M. Chauffe
St. Louis University

## Mark G. Jefferies
City of St. Louis Planning & Urban Design Agency

JONES & BARTLETT
LEARNING

*World Headquarters*
Jones & Bartlett Learning
5 Wall Street
Burlington, MA 01803
978-443-5000
info@jblearning.com
www.jblearning.com

Jones & Bartlett Learning books and products are available through most bookstores
and online booksellers. To contact Jones & Bartlett Learning directly, call 800-832-0034,
fax 978-443-8000, or visit our website www.jblearning.com.

Substantial discounts on bulk quantities of Jones & Bartlett Learning publications are available
to corporations, professional associations, and other qualified organizations. For details and
specific discount information, contact the special sales department at Jones & Bartlett Learning
via the above contact information or send an email to specialsales@jblearning.com.

ISBN: 978-1-4496-9860-7

Cover photo: © Stock Connection Distribution/Alamy Images

Printed in the United States of America
16 15 14 13 12     10 9 8 7 6 5 4 3 2 1

# Contents

# Preface

Most oceanography laboratory manuals require expensive equipment or primarily consist of tedious paper work. Some manuals additionally require students to be knowledgeable in chemistry, physics and geology. For these reasons, most schools do not offer an oceanography laboratory course to accompany lecture. We have attempted to address these concerns in three ways:

1. Laboratory exercises have been designed around safe, readily available, inexpensive and reusable materials.

2. Most of the laboratories are group-based activities that demonstrate principles discussed in lecture.

3. Laboratories only require minimal knowledge of science and math.

All of the exercises, except for three, are based upon hypothetical locations with no geographical basis. The latitudes and longitudes on the charts do not correlate with any known location. Of the three real areas cited in the text, two are in the Pacific Ocean and one in in the Atlantic.

Karl M. Chauffe
Department of Earth and Atmospheric Sciences
St. Louis University

Mark G. Jefferies
City of St. Louis Planning & Urban Design Agency

# Laboratory 1

# Metric-English Conversion, Understanding Graphs and the Graticule

## 1-1. THE METRIC SYSTEM

The French developed the metric system during the 1790s. Unlike the "English" system, in which the difference between smaller and larger units is irregular and has no scientific foundation, the metric system is based on multiples of 10. Smaller and larger units can be obtained simply by moving the decimal point, i.e. multiplying or dividing by 10. For example, in the "English" system there are 12 in to a foot, but 3 ft to a yard, and 1760 yards in a mile. In contrast, in the metric system there are 10 m to a dekameter, 10 dm to a hectometer, and 10 hm to a kilometer. The convenience and regularity of the metric system make it the standard system used in scientific research.

The basic units of the metric system are the **meter** for length, **gram** for mass (weight), **liter** for volume, and degree **Celsius** (°C) for temperature. The meter is equal to one ten-millionth of the distance between the North Pole and the equator. The gram is defined as the mass of one cubic centimeter (one millionth of a cubic meter) of water at 4°C. The liter is the volume of a cubic decimeter (one thousandth of a cubic meter). A degree Celsius is one-hundredth of the change in temperature between the freezing and boiling points of water. The French also experimented with decimal time, developing two types of clocks, one with 10 hours in a day and the other with one 100 hours in a day, but neither was widely accepted and both were eventually abandoned. The metric and English systems use the same divisions of time and angles.

Because all parts of the metric system use the same prefixes to indicate size, it is useful to learn them and their meanings. Some of these terms will be familiar to you because they are used to describe the size of computer memory, as in *kilo*byte (1000 bytes), *mega*byte (1 million bytes) and *giga*byte (1 billion bytes), or are parts of words with which you may be familiar, as in *kilo*watts (1000 watts), and in the names of insects, as in *milli*pede (1000 legs) and *centi*pede (100 legs).

| Prefix | | Size | | When combined with another word |
|---|---|---|---|---|
| giga | (G) | 1,000,000,000 | (1 billion) | times the size of 1 unit |
| mega | (M) | 1,000,000 | (1 million) | times the size of 1 unit |
| kilo | (k) | 1,000 | (1 thousand) | times the size of 1 unit |
| hecto | (h) | 100 | (1 hundred) | times the size of 1 unit |
| deka | (da) | 10 | (ten) | times the size of 1 unit |
| deci | (d) | 0.1 | (1 tenth) | times the size of 1 unit |
| centi | (c) | 0.01 | (1 hundredth) | times the size of 1 unit |
| milli | (m) | 0.001 | (1 thousandth) | times the size of 1 unit |
| micro | (μ) | 0.000001 | (1 millionth) | times the size of 1 unit |
| nano | (n) | 0.000000001 | (1 billionth) | times the size of 1 unit |

Until the U.S. adopts the metric system, as has most of the world and all branches of science, it will be necessary to learn how to convert units between the metric and English systems. Conversion factors are provided in a chart at the end of this laboratory. When converting between the English and metric systems, it is important to keep track of the units. Below are a few examples of how to convert units.

Example 1. Convert 15.7 mi into kilometers
   a. From the conversion chart at the end of Laboratory 1, **1 mi = 1.609 km**.
   b. This equation can be converted into a fraction equal to 1 either by dividing both sides by 1 mile or by dividing both sides by 1.609 km. So,

$$1 = 1.609 \text{ km}/1 \text{ mi} = 1 \text{ mi}/1.609 \text{ km}$$

   c. Any number multiplied by 1 is the same number. 15.7 mi × 1 is still 15.7 mi.
   d. Select the fraction equal to 1 that, when multiplied by 15.7 mi will cancel the miles and leave kilometers: 1 = 1.609 km/1 mi.
   e. If we substitute 1.609 km/1 mi for the number 1 in the equation 15.7 mi × 1, the miles will cancel and it will give the answer in kilometers:

$$15.7 \text{ mi} \times 1 = 15.7 \text{ mi.}$$
$$15.7 \cancel{\text{mi}} \times 1.609 \text{ km}/1 \cancel{\text{mi}} = 25.26 \text{ km}$$

Example 2. Convert 8.54 mi into centimeters.
   a. From Example 1 we already know that to convert miles to kilometers we multiply the number of miles by 1.609 km/mi. The conversion from kilometers to centimeters is similar.
   b. From the chart, 1 km = 1000 m and 1 m = 100 cm.
   c. The equation 1 km = 1000 m can be converted into a fraction equal to 1 either by dividing both sides by 1 km or by 1000 m. Similarly, the equation 1 m = 100 cm can be converted into a fraction equal to 1 by dividing both sides by 1 m or 100 cm.
   d. Thus, 1 = 1 km/1000 m = 1000 m/1 km  and  1 = 1 m/100 cm = 100 cm/1 m.
   e. Any number multiplied by 1, no matter how many times, is still the same number:

$$8.54 \text{ mi} \times 1 \times 1 \times 1 = 8.54 \text{ mi.}$$

   f. Select the fractions equal to 1 that, when multiplied by 8.54 mi, will cancel miles, kilometers and meters and leave centimeters.

   g. If we substitute 1.609 km/1 mi, 1000 m/1 km and 100 cm/1 m into the equation, the units miles, kilometers and meters cancel and the answer is in centimeters: 8.54 mi × 1 × 1 × 1 = 8.54 mi.

$$8.54 \ \cancel{mi} \times 1.609 \ \cancel{km}/1 \ \cancel{mi} \times 1000 \ \cancel{m}/1 \ \cancel{km} \times 100 \ cm/1 \ \cancel{m} = 1{,}374{,}086 \ cm.$$

**Example 3. Convert 16 km into inches.**

   a. From the chart we know 1 km = 0.6214 mi, 1 mi = 5280 ft and 1 ft = 12 in.

   b. Thus, 1 = 0.6214 mi/km = 1 km/0.6214 mi, 1 = 5280 ft/mi = 1 mi/5280 ft and 1 = 12 in/ft = 1 ft/12 in.

   c. Any number multiplied by 1, no matter how many times, is still the same number:

$$16 \ km \times 1 \times 1 \times 1 = 16 \ km.$$

   d. Select the fractions equal to 1 that, when multiplied by 16 km, will cancel kilometers, miles, and feet and leave inches:

$$16 \ km \times 1 \times 1 \times 1 = 16 \ km.$$
$$16 \ \cancel{km} \times 0.614 \ \cancel{mi}/\cancel{km} \times 5280 \ \cancel{ft}/\cancel{mi} \times 12 \ in/\cancel{ft} = 622{,}448.64 \ in.$$

**Example 4. Convert 1.245 ft³ into liters.**

   a. From the chart we know 1 ft³ = 1728 in³, 1 in³ = 16.39 cm³, 1 cm³ = 1 ml and 1 L = 1000 ml

   b. Thus, 1 = 1728 in³/ft³ = 1 ft³/1728 ft³, 1 = 16.39 cm³/in³ = 1 in³/16.39 cm³,

$$1 = 1 \ cm^3/ml = 1 \ ml/cm^3 \quad \text{and} \quad 1 = 1 \ L/1000 \ ml = 1000 \ ml/1 \ L.$$

   c. Any number multiplied by 1, no matter how many times, is still the same number:

$$1.245 \ ft^3 \times 1 \times 1 \times 1 \times 1 = 1.245 \ ft^3.$$

   d. Select the fractions equal to 1 that, when multiplied by 1245 ft³, will cancel cubic feet, cubic inches, cubic centimeters and millimeters and leave liters.

$$1.245 \ ft^3 \times 1 \times 1 \times 1 \times 1 = 1.245 \ ft^3.$$
$$1.245 \ \cancel{ft^3} \times 1728 \ \cancel{in^3}/\cancel{ft^3} \times 16.39 \ \cancel{cm^3}/\cancel{in^3} \times 1 \ \cancel{ml}/\cancel{cm^3} \times 1 \ L/1000 \ \cancel{ml} = 35.26 \ L.$$

# EXERCISE 1.  CONVERSION OF UNITS

1. Convert each of the following into the units requested.

   a. Earth's circumference, 24,900 mi   _____ km   _____ mi.

   b. Equatorial diameter, 12,756 km   _____ ft   _____ mi

   c. Ocean depth of 36,198 ft   _____ m   _____ km

   d. Highest point on Earth, 8847 m   _____ km   _____ ft

   e. Wind speed of 30 mi/h   _____ km/hr   _____ ft/min

f.  Density of 60 lbs/ft$^3$ _____ kg/m$^3$ _____ g/cm$^3$

g.  Density of 5.8 g/cm$^3$ _____ kg/m$^3$ _____ lbs/in$^3$

h.  Speed of Earth's equatorial
    rotation, 460 m/sec _____ mi/hr _____ km/hr

i.  Discharge of the Mississippi
    River, 61,000 ft$^3$/sec _____ m$^3$/sec _____ l/sec

j.  2500 mi$^2$ _____ km$^2$ _____ m$^2$

k.  2500 km$^2$ _____ ft$^2$ _____ mi$^2$

l.  Mean density of the Earth,
    5.522 g/cm$^3$ _____ lbs/ft$^3$ _____ kg/m$^3$

m.  10 L container _____ Gal _____ in$^3$

n.  55 gallon drum of oil _____ ft$^3$ _____ liters

## 1-2. GRAPHS

The term "datum" refers to one unit of information. The plural of datum is "data." In science data typically consist of sets of numbers. Because it is difficult to comprehend a large series of numbers and to identify trends within them, other methods of presenting data are frequently used. A **graph** is a pictorial representation of a series of numbers. It allows complex information to be presented in a format that is easily understood and readily shows trends, relationships and changes. There are many types of graphs. Those with which you are most likely to be familiar are the pie diagram, bar diagram, and line graph.

On the **pie diagram** (Figure 1-1), segments of a circle are used to represent percentages of the whole. **Bar diagrams** (Figure 1-2) consist of one or more horizontal or vertical bars plotted against a scaled axis. The longer the bar, the larger the amount the bar represents. A thermometer is a very simple type of bar scale, where the height of the mercury (the bar) is plotted against the units of temperature (the scale) at a moment in time.

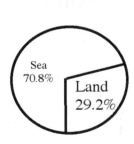

Figure 1-1. Pie diagram of the percentage of Earth's surface that is land and sea.

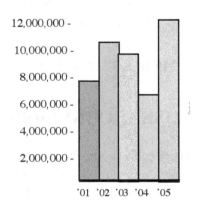

Figure 1-2. Bar diagram of tons of fish caught per year.

In oceanography and many other areas of science, the most commonly used graph is the **line graph**. It typically consists of two variables plotted on scales arranged perpendicularly to each other. The horizontal scale is called the **x-axis** and the vertical scale is called the **y-axis**. When a number of data points have been plotted, based on their x and y coordinates, they are connected by a line (Figure 1-3). Configuration of the line reveals trends and how the variables are related. A **direct relationship** between the variables means that they both increase or decrease together. For example, on Figure 1-3, location A, as depth increased, the number of fish observed also increased.

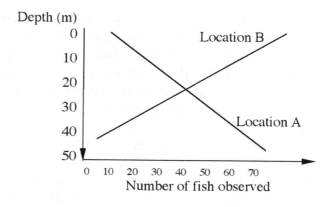

Figure 1-3. Graph plotting number of fish observed at various depths at locations A and B. At A there is a direct relationship between the number of fish and depth; the number of fish observed increases as depth increases. At B there is an inverse relationship between the number of fish and depth; the number of fish observed decreases as depth increases.

An **inverse relationship** means that as one variable increases, the other variable decreases. Figure 1-3 indicates that at location B, the number of fish observed increased as depth decreased. Frequently, a relationship can be much more complex and the line is irregularly curved. On graphs used in oceanography, the y-axis is commonly depth and the x-axis is a **parameter** (**variable**) measured against depth, such as temperature, oxygen content, salinity, pressure, velocity of sound, or water density, etc. An important aspect of reading any line graph is understanding the scales used along the x- and y-axes. On some graphs, the scales are **linear** and the distance between units is uniform (see Figure 1-5). On other graphs, one or both scales may be **nonlinear** and the distance between equivalent units varies (see Figure 1-7). A logarithmic scale is an example of a non-linear scale because the distance between units of equal size changes along the scale. Always examine the scales and be certain you understand them before attempting to interpret a graph.

When comparing two graphs using the same units and scale type, be certain that the spacing between units on the two graphs is identical. For example, if the units are 1 cm apart on one graph but 2 cm apart on the other, the scales are not identical. Changing the spacing can drastically alter the slope and shape of a line, making it appear steeper or gentler (Figure 1-4). Misinterpretation of the graphs will result if the difference in the scales in not recognized.

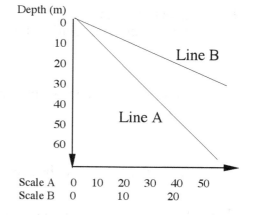

Figure 1-4. Same data plotted at different horizontal scales. Line A scale A; line B scale B. Note the difference in line slope.

**Logarithmic scales** are frequently used when the data encompass a very wide range of values. For example, if the data range from 1 to 1,000,000, the graph would have to be enormously large for all of the data to be presented. In contrast, on a logarithmic scale all the data could be easily accommodated and the size of the graph would not be excessive. A **logarithmic scale** represents a series of equal units, but the spacing decreases with increasing value within a **cycle** of 10 units. On the Depth scale of Figure 1-6, depths are indicated in 1 m increments, but the distance between the lines representing 1–2 m depth is much greater than that between 9–10 m depth. If two cycles are used on the graph, the second cycle is identical in spacing to the first, but its units are assigned 10 times the value of the first cycle. On Figure 1-6, the first cycle is for depths 1–10 m, the second cycle 10–100 m and the third cycle 100–1000 m. To graph numbers from 1 to 1,000,000 would require only six cycles. Note that logarithmic scales do not start with zero.

## EXERCISE 2.   COMPARISON OF GRAPHS

The following questions refer to the data recorded on the change in sea water temperature with depth.

| Depth (in m) | Temperature (in °C) | Depth (in m) | Temperature (m °C) | Depth (in m) | Temperature (m °C) |
|---|---|---|---|---|---|
| 0 | 32.0 | 11 | 26.0 | 40 | 03.2 |
| 1 | 31.0 | 12 | 24.25 | 50 | 03.0 |
| 2 | 30.5 | 13 | 20.5 | 60 | 02.9 |
| 3 | 30.25 | 14 | 19.5 | 100 | 02.7 |
| 4 | 30.0 | 15 | 17.0 | 200 | 02.6 |
| 5 | 29.75 | 16 | 14.2 | 300 | 02.5 |
| 6 | 29.40 | 17 | 11.3 | 400 | 02.4 |
| 7 | 29.15 | 18 | 09.0 | 700 | 02.3 |
| 8 | 29.03 | 19 | 07.0 | 800 | 02.2 |
| 9 | 29.0 | 20 | 05.75 | 900 | 02.1 |
| 10 | 28.5 | 30 | 03.4 | 1000 | 02.0 |

## Questions

1. From the data above:
   a.  What is the general relationship (direct or inverse) between temperature and depth?

   b.  Is this relationship consistent and uniform throughout? If not, explain how it varies.

   c.  What trends, if any, can be seen in data?

2. Plot the data on linear graph paper (Figure 1-5), semi-log graph paper (Figure 1-6), and log-log graph paper (Figure 1-7) and answer the questions that follow. (*Note*: a depth of zero can not be plotted on logarithmic scales. The scale begins at a depth of

1 m and has the first cycle extend from 1 to 10 m, the second cycle from 10 to 100 m, and the third cycle from 100 to 1000 m.)

a.  What is the general relationship (direct or inverse) between temperature and depth?

b.  Is this relationship consistent and uniform throughout?

c.  What trends, if any, can be seen in data?

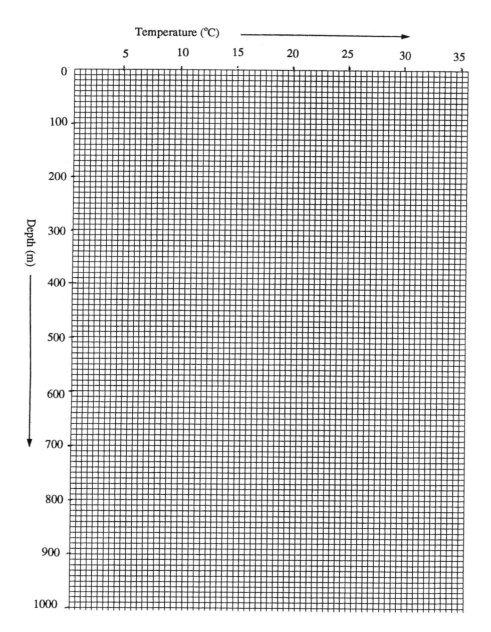

Figure 1-5.  Linear graph paper.

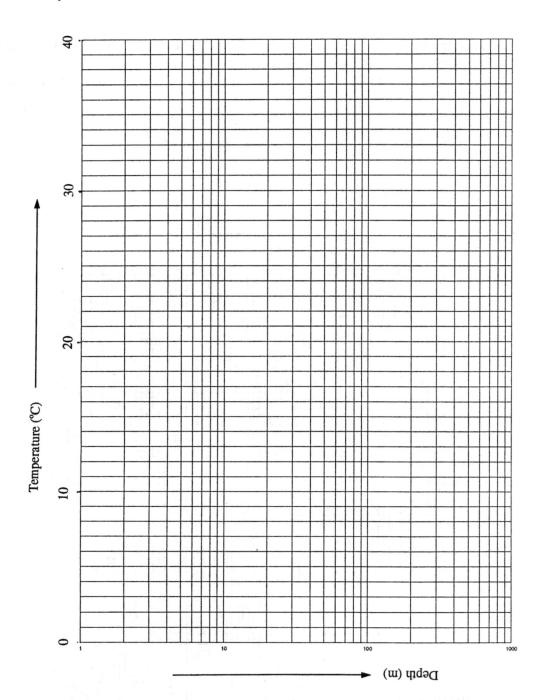

Figure 1-6.  Semi-log graph paper. Temperature is plotted on a linear scale and depth on a logarithmic scale. The paper is referred to as semi-log because only one scale is logarithmic.

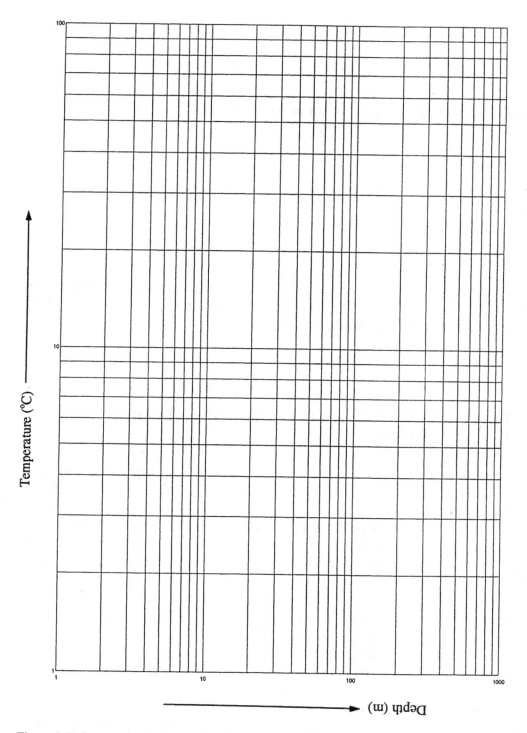

Figure 1-7.  Log-log graph paper. Both temperature and depth are plotted on logarithmic scales.

3. Were the data set or the graphs more revealing and easier to understand? Why?

4. Which graph best portrayed trends in the data? Why?

5. What are the disadvantages and advantages of presenting the data on each type of graph?

6. Why are logarithmic scales commonly used when one parameter varies by several magnitudes relative to the other? (For example, depth varied by 10,000 m but temperature varied only about 30°C.)

## 1-3.  THE GRATICULE: LATITUDE, LONGITUDE AND ANTIPODES

A **great circle** is formed by the intersection of the Earth's surface and any plane passing through Earth's center. All great circles are equal in length to Earth's circumference and an infinite number of great circles can be drawn on the Earth. Any two points on Earth's surface can be connected by only one great circle, unless the points are **antipodes**, points on the exact opposite sides of the Earth that can be connected by a straight line passing through Earth's center. An **arc** is a portion of a great circle. Arcs of great circles always are the shortest distance between two points on Earth.

A **small circle** is formed by the intersection of Earth's surface and any plane not passing through Earth's center. Small circles do not represent the shortest distance between two points. An infinite number of small circles can be drawn to connect any two points on the Earth's surface unless they are antipodes. Only a great circle can connect antipodes.

A **graticule** is a network of lines intersecting at right angles that can be used as a reference system for locating points. The graticule used to locate points on Earth's surface is based on great and small circles and consists of lines of latitude and longitude.

**Lines of latitude** are formed by the intersection of Earth's surface with planes passing perpendicularly through Earth's axis. Only one line of latitude, the equator, is a great circle; all others are small circles. Poleward from the equator, lines of latitude become progressively smaller circles and are reduced to a point at the poles. Lines of latitude are numbered based their angular distance north or south of the equator, as measured relative to the Earth's center (Figure 1-8). They are numbered consecutively north or south of the equator, with the equator being 0° and the poles being 90° north and south, respectively. The term **low latitude** refers to the area near the equator, **midlatitude** identifies the region around 45° and **high latitude** indicates polar regions. Lines of latitude are sometimes called **parallels** because they are all parallel to each other.

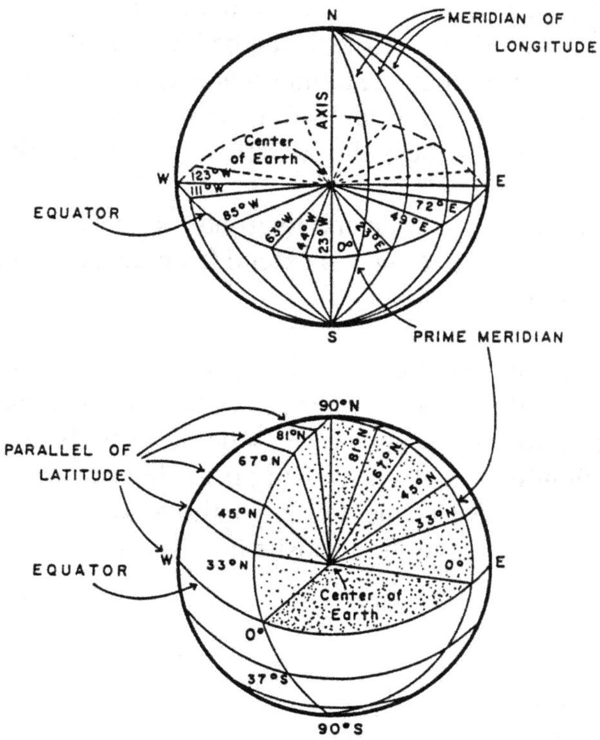

Figure 1-8. Latitude and longitude of the graticule. Upper figure shows angular divisions east and west of the Prime Meridian. Lower figure illustrates angular divisions north of the equator.

**Lines of longitude** or **meridians** are formed by the intersection of Earth's surface with planes passing through the length of Earth's axis. All lines of longitude are arcs of great circles and are equal to one-half Earth's circumference. They extend from the North to South Rotational Poles. Lines of longitudes are numbered based on their angular distance east or west of the **Prime Meridian** (Figure 1-8). The prime meridian is 0° and passes through Greenwich, England. To the east of the Prime Meridian, lines of longitude are measured in degrees up to 180° E. To the west, longitude is measured up to 180° W. The 180° meridian is also the **International Date Line**.

Latitude and longitude are usually listed along the borders of charts and maps. For most, latitude is indicated along the right and left sides and longitude is shown along the top and bottom. Because degrees are rather large units, smaller divisions are needed for more precise location. One degree is divided into 60 **min** (60′) and 1 min is divided into 60 **sec** (60″).

$$[1° = 60′; 1′ = 60″]$$

As used here, seconds and minutes are divisions of an angle and do not refer to time. Any point on Earth's surface can be located in terms of latitude and longitude by stating the number of degrees, minutes, and seconds north or south of the equator and the number of degrees, minutes, and seconds east or west of the Prime Meridian. For example, San Francisco is located at 37°49'N and 122°25'W.

The words latitude and longitude are easily confused. Remember that all lines of *long*itude are *long* and extend from the North to South Pole. Lines of latitude become shorter poleward.

The antipode for any point on Earth can be easily located using two cross sectional views of Earth: one axial (parallel to the axis) and one equatorial (parallel to the equator). For example, to locate the antipode of 38°50'N 90°05'W, begin with the axial cross section (Figure 1-9). A line passing from 38°50'N latitude through Earth's center would emerge at 38°50'S latitude. For any latitude, the latitude of the antipode is always the same number, but in the opposite hemisphere.

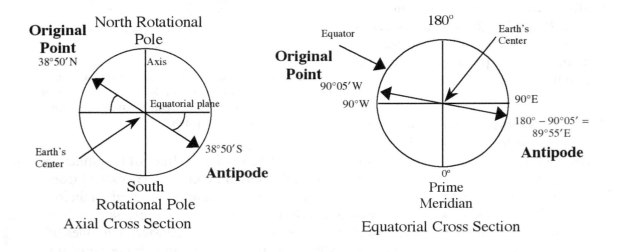

Figure 1-9. Axial and equatorial views showing the latitudinal relationship between a point and its antipode.

To determine the longitude of the antipode, use the equatorial cross section (Figure 1-9). However far the original point is east or west from the Prime Meridian, the antipode will be that far away from 180° in the opposite hemisphere. Subtract the longitude of the original point from 180° to determine the longitude of the antipode. For example, a line from 90°05'W passing through Earth's center would emerge on the opposite side of the Earth at 180–90°05' = 89°55'E.

## EXERCISE 3.   THE GRATICULE; ANTIPODES, LATITUDE AND LONGITUDE

1. If you were halfway between the equator and the South Pole, and a quarter of the way around the Earth west of the Prime Meridian, what would be your latitude and longitude?

   _____

2. You begin at a point 10°N 30°E and you move to a new location 25° to the south and 40° to the west of your original position. What is you new latitude and longitude?

   _____

3. Identify the antipode of each of the following.

   a.  90°N   45°W         _____

   b.  0°N   0°E           _____

   c.  56°N   38°E         _____

   d.  14°S   170°W        _____

4. Given the following locations, answer the questions below.

   | Los Angeles | 33°42′N | 118°15′W |
   |---|---|---|
   | San Francisco | 37°49′N | 122°25′W |
   | Hanna Bay, Hawaii | 20°45′N | 155°59′W |

   a.  What is the difference in latitude between Los Angeles and San Francisco?
   _____

   b.  What is the difference in latitude between San Francisco and Hanna Bay?
   _____

   c.  What is the difference in longitude between Los Angles and San Francisco?
   _____

5. Using a world map or globe, determine the latitude and longitude of the following as precisely as possible.

   a.  Washington D.C.      _____

   b.  Sydney, Australia    _____

   c.  Warsaw, Poland       _____

   d.  Lima, Peru           _____

6. On the hemisphere below, draw the following significant lines of latitude and label them. These lines of latitude will be referenced in later labs.

a. Equator
b. North and South Poles
c. 30°N and 30°S
d. 60°N and 60°S
e. 23.5°N, Tropic of Cancer

f. 23.5°S, Tropic of Capricorn
g. 66.5°N, Arctic Circle
h. 66.5°S, Antarctic Circle

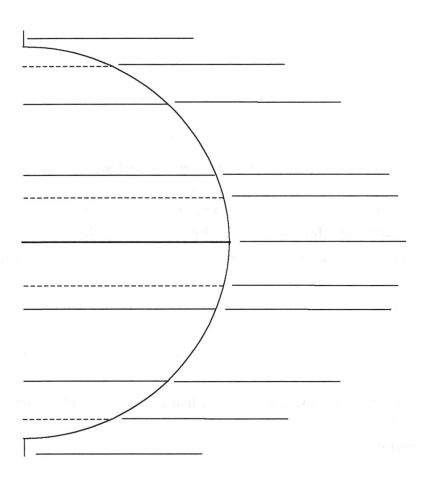

# Conversion Tables for Metric and English Systems

## LINEAR MEASUREMENTS

*English*
12 inches = 1 foot
3 feet = 1 yard = 36 inches
1 mile = 1760 yards = 5280 feet

*Metric*
1 micron = 0.001 millimeter
10 millimeters = 1 centimeter
100 centimeters = 1 meter
1000 meters = 1 kilometer

## AREA MEASUREMENTS

*English*
1 square foot = 144 square inches
1 square yard  = 9 square feet
1 square mile = 27,878,400 square feet

*Metric*
1 square meter = 10,000 square centimeters
1 square kilometer = 1,000,000 square meters

## VOLUMETRIC MEASUREMENTS

*English*
1 quart  = 2 pints = 57.75 cubic inches
4 quarts  = 1 gallon = 231 cubic inches
1 cubic foot = 1728 cubic inches = 7.48 gallons
1 cubic yard = 27 cubic feet = 201.97 gallons
1 cubic inch = 0.554 fluid ounce

*Metric*
1 liter = 1000 milliliters = 0.001 cubic meter
1 milliliter = 1 cubic centimeter
1 cubic meter = 1,000,000 cubic centimeters

## WEIGHT and MASS

*English*
1 pound = 16 ounces
1 ton = 2000 pounds

*Metric*
1000 grams = 1 kilogram
1000 kilograms = 1 metric ton

## TIME DIVISIONS

60 seconds (sec) = 1 minute (min)
60 minutes (min) = 1 hour (hr)
24 hours (hr) = 1 day

### Angular Divisions*

60 seconds (″ or sec) = 1 minute (′ or min)
60 minutes (′ or min) = 1 degree (°)

(*Note. Angular divisions of second and minute are not
related to the time units of seconds and minutes.)

## ENGLISH-METRIC CONVERSION

*Linear*
1 inch (in) = 2.54 centimeters (cm)
1 foot (ft) = 0.3048 meter (m)
1 yard (yd) = 0.9144 meter (m)
1 mile (mi) = 1.609 kilometers (km)

*Volume*
1 cubic inch (in$^3$) = 16.39 cubic centimeters (cm$^3$)
1 cubic foot (ft$^3$) = 0.0283 cubic meter (m$^3$)
1 cubic yard (yd$^3$) = 0.7646 cubic meter (m$^3$)
1 quart (liquid) = 0.946 liter (l)
1 gallon (US) = 0.003785 cubic meter (m$^3$)

*Area*
1 square inch (in$^2$) = 6.452 square centimeters (cm$^2$)
1 square foot (ft$^2$) = 0.0929 square meter (m$^2$)
1 square yard (yd$^2$) = 0.836 square meter (m$^2$)
1 square mile (mi$^2$) = 2.59 square kilometers (km$^2$)

*Weight and Mass*
1 pound (lb) = 0.4536 kilogram (kg)
1 ton = 907.2 kilograms (kg)

## METRIC-ENGLISH CONVERSIONS

*Linear*

1 millimeter (mm) = 0.0394 inch (in)

1 centimeter (cm) = 0.394 inch (in)

1 meter (m) = 3.281 feet (ft) = 1.904 yards (yd)

1 kilometer (km) = 0.6214 mile (mi)

*Volume*

1 cubic centimeter ($cm^3$) = 0.061 cubic inch ($in^3$)

1 cubic meter ($m^3$) = 35.3 cubic feet ($ft^3$)

1 liter (l) = 1.057 quarts (qt)

1 cubic meter ($m^3$) = 264.2 gallons (gal)

*Area*

1 square centimeter ($cm^2$) = 0.16 square inch ($in^2$)

1 square meter ($m^2$) = 10.764 square feet ($ft^2$)

1 square kilometer ($km^2$) = 0.3861 square mile ($mi^2$)

*Weight and Mass*

1 kilogram (kg) = 2.2 pounds (lbs)

1 metric ton = 2205 pounds (lbs)

# Laboratory 2

# Bathymetric Charts and Scientific Notation

## 2-1. BATHYMETRIC CHARTS

Nautical **charts** are a two-dimensional representation of a portion of a body of water at a reduced size. They are primarily used for navigation and piloting. **Bathymetric charts** indicate depth and display the **bathymetry** (elevations and depressions) of the sea floor. **Bottom sediment charts** show the distribution and type of sediment or rock exposed on the sea floor. Most navigation charts indicate both bathymetry (depth) and bottom composition.

A **sounding** is the determination of depth to the sea floor. Initially, soundings were obtained by lowering a weighted, measured cable to the bottom. Today, soundings are obtained with a **sonic depth recorder**. This device uses sound waves to determine depth. The method is faster, more accurate and allows continuous depth determination as a ship travels (Figure 2-1). A **sound generator** on the ship emits sound waves that strike the sea floor and are reflected upward to a **hydrophone**, a listening device. Shipboard computers record the round-trip time of the sound waves and calculate depth by multiplying the speed of sound in water ($S_w \cong 1460$ m/sec) by half of the travel time:

$$D = S_w \, x \, \tfrac{1}{2} \, time$$

Half of the travel time is the travel time to or from the bottom and is equal to the depth of the water. For example, if the total travel time was 4 sec, time to reach

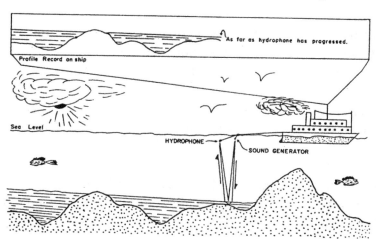

Figure 2-1. Sonic depth recorder recording a continuous depth profile.

bottom was only 2 sec. Two seconds times the speed of sound in water ($S_w$) equals 2920 m (2 ~~sec~~ × 1460 m/~~sec~~), the depth of the sea at that point. On charts, soundings are shown as black or blue numbers. Frequently, they form the basis for contoured bathymetric charts.

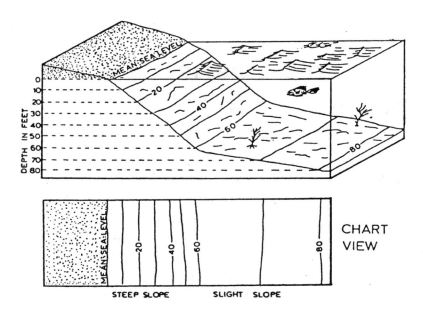

A **contour line** is a line on a chart that connects points of equal depth. For example, the −20 ft contour line connects points 20 ft below sea level (Figure 2-2). The −40 ft contour connects points 40 ft below sea level. The **datum** (base level or zero contour) for most charts is **mean sea level**, the average sea level, ignoring high and low tides. The change in depth between two adjacent contours is the **contour interval** For Figure 2-2, the contour interval is 10 ft. Darker, numbered contour lines on charts are called **index contours** The number printed in the gap in the line is the depth represented by the line. Unnumbered lines between index contours are called **supplemental contours** As with soundings, contour lines may be printed in blue or black.

Figure 2-2. Contour lines and their representation on charts. The closer together the contours the steeper the slope.

Although contours are not very accurate, except where they coincide with a sounding, they provide a useful bathymetric "picture" of the seascape not easily seen from soundings alone. Contours outline submarine hills, valleys, ridges, and mountains.

In addition to soundings and contours, depth may also be indicated by color. On official U.S. government charts, water 0–18 ft deep is colored dark blue, 18–36 ft light blue, and deeper than 36 ft very light blue to white.

Depths on charts may be expressed in feet, meters or **fathoms** A general relationship among these units is: 6 ft ≅ 2 m = 1 fathom. To determine which units are used on a given chart, consult the chart's **legend** an area on the chart explaining what various symbols mean. On Figure 2-8, the legend is in the lower right-hand corner.

## 2-2.  CONTOURING A BATHYMETRIC CHART

In constructing a contour bathymetric chart from soundings, the following guidelines and information should be considered.

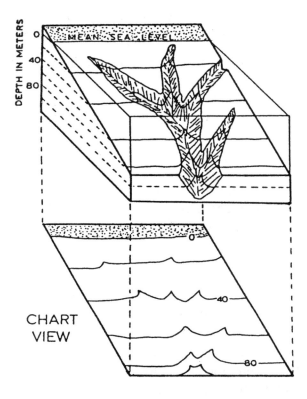

Figure 2-3.  Where contours cross a valley they form Vs that point upslope.

1. Contour lines only connect points of equal depth. They do not extend uphill or downhill to greater or lesser depths. Contour lines are usually smoothly curving lines.

2. Where contours cross valleys, they form a distinct "V" pointing upvalley. Because valleys are eroded into a slope and because contours must remain at the same depth on the bottom, contour lines inscribe a V within a valley with the point of the V directed upvalley (Figure 2-3).

3. Contour lines always close. They form regular to highly irregular circles or loops. If the complete contour is not contained on the chart, the contour line will terminate abruptly at the edge of the chart move back, as on Figure 2-3.

4. Contour lines never split or cross, but may merge at overhanging cliffs or vertical slopes (Figure 2-4). Contour lines enclose areas of equal or greater elevation. Thus, it is not necessary to connect every point of the same elevation with a contour if the contour encloses those points. If contours crossed, it would imply that the point of intersection is at two depths simultaneously, an impossibility.

5. **Slope** or **gradient** refers to the inclination of the sea floor or land surface. Steepness of slope is indicated by the distance between contours. The closer together the contours, the steeper the slope. The farther apart the contours, the gentler the slope (see Figure 2-2).

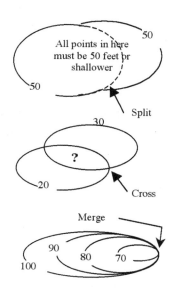

Figure 2-4.  Contours never split or cross, but may merge.

6. Normally, contour lines enclose areas of shallower depth. For a local, closed depression, where contours enclose an area of greater depth, the contours are distinguished by small tick marks **(hachured marks)** that point into the depression (Figure 2-5). These are called **depression contours**.

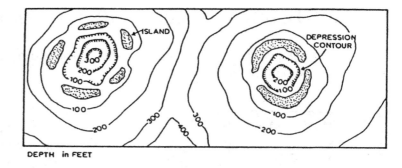

Figure 2-5. Depression contours are indicated by hachured marks (tick marks pointing downslope) on the contour lines.

7. **Depth** is the term for measurements below sea level. Measurements above sea level are called **Elevation**. **Relief** refers to the change in depth or elevation between two points. For example, if point A is 30 m deep and point B is 45 m deep, the relief between them is 15 m.

## 2-3.   CHART SCALE AND HORIZONTAL DISTANCE

Because charts represent the Earth's surface at a reduced size, it is necessary to know the amount of reduction to be able to correctly interpret the chart. **Scale** is the fixed relationship between a distance on the chart and the corresponding distance on the Earth. Scale may be indicated in several ways. **Representative fraction** is a ratio or fraction relating distances measured on a chart to the same distance on the Earth's surface. No units are specified. For example, the representative fraction 1/125,000 and the ratio 1:125,000 state that any 1 unit distance on the chart equals 125,000 of the same units on the Earth's surface. One centimeter on the chart equals 125,000 cm, 1 in on the chart equals 125,000 in, and even one paperclip on the charts equals the length of 125,000 paperclips on the ground. In other words, if you reduced Earth's surface 125,000 times, the area illustrated by the chart would be the same size as the chart.

A **graphic bar scale** is a bar or line subdivided into convenient Earth surface distances. A length measured on the chart can be compared to the bar scale and the corresponding Earth surface distance read directly. Always note where the zero is on the bar scale. For Figure 2-6, the total length of the bar scale is four miles.

1        0        1        2        3 Km

Figure 2-6. Graphic bar scale. Note 0 position.

**Stated ratios** are similar to representative fractions, but specific units of length are assigned to the ratio. For example, 1 cm = 2 km means 1 cm on the chart is equivalent to 2 km on Earth's surface.

If a chart is photographically enlarged or reduced, the stated ratio and representative fractions are no longer valid. The bar scale will still read true if its size is altered proportionally to the chart.

Representative fraction, graphic bar scale and stated ratio for a given chart all provide the same information about the chart. If the bar scale is divided so that a length of 1 cm is designated as 2 km, then the representative fraction is 1/200,000 (1 km = 100,000 cm) and the stated ratio is: 1 cm = 2 km.

## 2-4. DETERMINING SLOPE OR GRADIENT

**Slope** of the sea floor, also called **gradient**, may be numerically expressed as a ratio, percentage or in degrees. As a ratio, it is equal to the **relief** (change in depth) divided by the horizontal distance over which the slope is measured. Slope normally is stated in feet/mile, meter/kilometer, fathom/mile or fathom/kilometer.

**Slope or Gradient = relief/horizontal distance of slope**

**Percentage slope** is equal to relief divided by horizontal distance multiplied by 100%. [NOTE: Both relief and distance must be expressed in the same units so that units are eliminated from the answer.]

**% Slope = (relief/horizontal distance of change) × 100%**

For example, if the slope is 100 ft/mi, percent slope would be:

**(100 ft/mi × 1 mi/5280 ft) ×
100% = 1.9%.**

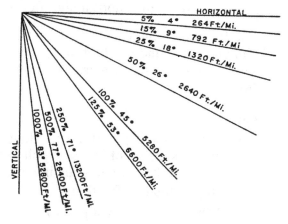

Calculating slope (gradient) as **degrees of slope** is complicated and will not be discussed herein, but the general relationship between percent slope, degree slope and gradient is illustrated in Figure 2-7. It is interesting to note that 100% slope is a 45° slope. A horizontal line is 0% slope and 0° slope. A vertical line is a 90° slope and an infinity percent slope.

Figure 2-7. Slope given as percent slope, degrees, and feet per mile.

## EXERCISE 1.  CONTOURING

1. Sandy Harbor chart
   Convert the sounding chart of Sandy Harbor (Figure 2-8) into a contoured bathymetric chart using a contour interval of 1 fathom.

   Part of the 1 fathom contour line has been drawn. Location of this line was determined by comparing pairs of soundings. Beginning on the upper left side of the chart, between soundings points 0.1 and 1.9 fathoms, there must be some point 1fathom deep and it is probably midway between these points. Continuing to the right, between soundings 0.5 and 1.9, and between 2.7 and 1.8, there would have to be points 1 fathom deep. After determining all probable points 1 fathom deep, the 1 fathom contour line can be drawn as a smoothly curving line. Once you have drawn one or two contour lines using this depth-comparison method, you will discover that it is not necessary to compare all points and contouring will proceed much faster.

   There are many different, but correct, charts that can be produced from the same soundings, if you follow the rules. Should your finished chart not be identical to your classmate's, it does not necessarily mean that either of you are in error.

   Questions
   1. What is the depth at point A in fathoms? _____ meters? _____
   2. Where is the deepest part of the bay? _____
   3. What is the relief between points A and B? _____
   4. Determine the slope of the bay from points A to B as indicated below:
      _____fathom/mi _____% _____ft/mi _____m/km
   5. What is the representative fraction for this chart? _____
   6. What is the stated ratio for this chart? _____
   7. Six inches measured on this chart equals how many feet on the Earth? _____
   8. What is the relationship between representative fraction and stated ratio for this chart?
   9. If the chart had been contoured using meters or yards as the contour interval, would the map appear significantly different?

2. Pacific Ocean Chart
   Convert the sounding chart of a portion of the southern Pacific Ocean (Figure 2-9) into a contoured bathymetric chart. Draw contours for 100 m, 200 m, 400 m, 600 m, etc.

   Questions
   1. What is the depth of the sea floor at point A? _____ B? _____
   2. What is the relief between points A and B? _____
   3. Convert the graphic scale into a representative fraction and stated ratio.
      Representative fraction 1: _____ Stated ratio 1 in = _____
   4. What is the depth at point Z? _____

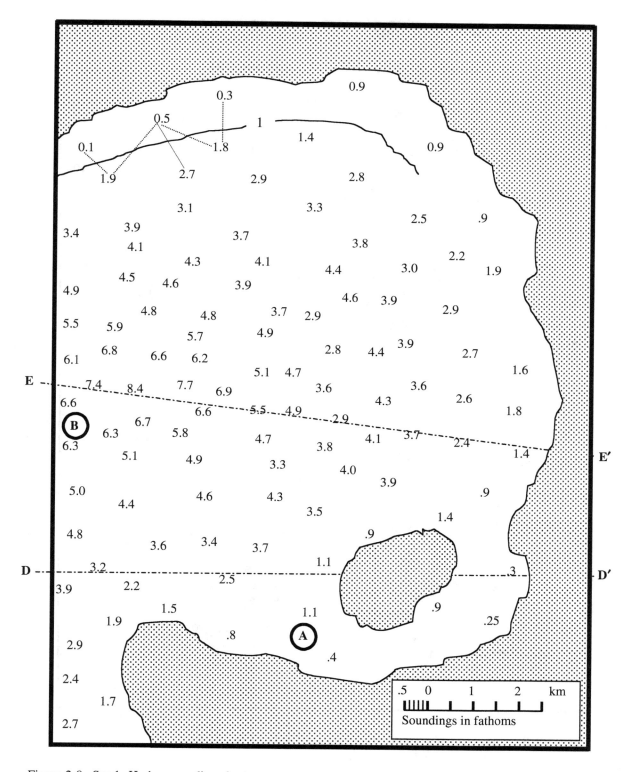

Figure 2-8. Sandy Harbor sounding chart.

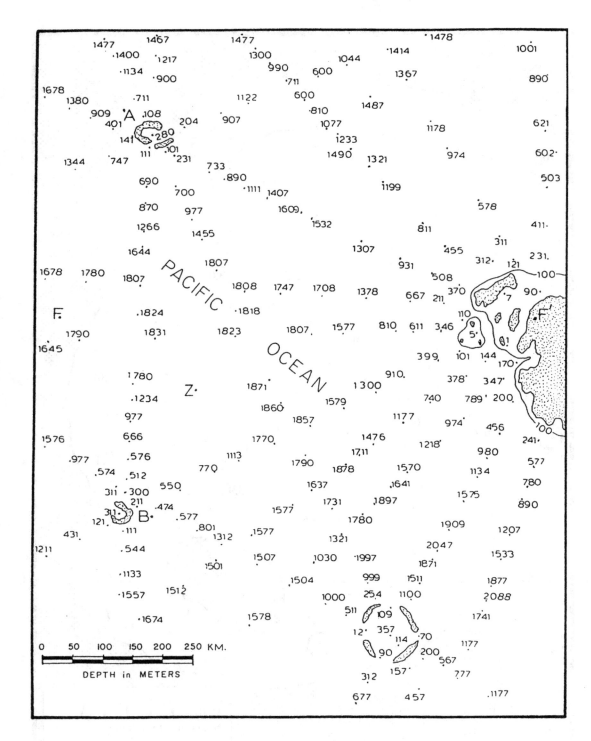

Figure 2-9. Sounding chart of a portion of the southern Pacific Ocean.

## 2-5. BATHYMETRIC PROFILES

A **bathymetric profile** is a vertical section that provides a "skyline view" of the sea floor along a line. It graphically illustrates, in silhouette form, hills as rises and valleys as depressions.

For a profile to accurately illustrate the sea floor, a ratio of 1:1 for vertical and horizontal distances is required. This means that one unit on the vertical scale is represented the same as one unit on the horizontal scale. Because relief is generally small and horizontal distance great, this is rarely possible. For example, a profile across the Atlantic Ocean extends for thousands of kilometers, but the total relief is only a few kilometers. If displayed with a 1:1 ratio on a regular sheet of paper, the profile would appear as a flat line. Thus, it is common to exaggerate (stretch) relief.

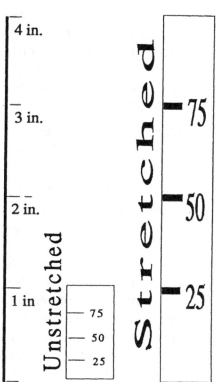

**Vertical exaggeration** (**VE**) is the number of times by which the vertical scale is stretched. For example, if vertical exaggeration is ×4, this means that the vertical scale has been stretched four times its original length. This can be easily demonstrated with a rubber band. On an unstretched rubber band, measure and mark quarter inches for the distance of an inch. If we use a scale of 1 in equals 100 ft, each quarter inch equals 25 ft and the total length (1 in) represents 100 ft (Figure 2-10). Stretch the rubber band until the original inch is 4 in long. If you measure along the stretched rubber band you will discover that 1 in now represents only 25 ft. Distances have been stretched by a factor of four.

For most profiles there are two scales: one for the horizontal and an exaggerated one for the vertical. In the above example, the horizontal scale would be 1 in = 100 ft and the vertical scale is 1 in = 25 ft. To determine the vertical exaggeration, simply divide the horizontal stated ratio by the vertical stated ratio:

Figure 2-10.  Vertical exaggeration on a rubber band.

### Vertical Exaggeration = Horizontal Stated Ratio/Vertical Stated Ratio

$$VE = (1 \text{ in} = 100 \text{ ft})/(1 \text{ in} = 25 \text{ ft}) = 4$$

Vertical exaggeration causes distortion and the amount of distortion increases with the amount of exaggeration. On a wide rubber band, draw a circle. As the rubber band is stretched, the circle becomes distorted into an oval and eventually an ellipse. As vertical exaggeration increases on a profile, hills appear to be higher, valleys deeper and the slopes between them become much steeper. Gentle slopes will look steep; steep slopes will seem precipitous. Vertical exaggeration does not alter flat areas nor does it steepen a vertical slope, but it will make the vertical drop appear to be greater.

## 2-6. CONSTRUCTION OF A BATHYMETRIC PROFILE

To construct a bathymetric profile, several steps must be followed (Figure 2-11).

1.  Choose the line, called a **trace,** on the chart along which the profile is to be constructed. Determine the contour interval for the map and calculate the relief along the trace by subtracting the value of the shallowest contour from the deepest contour that crosses the trace.

2.  Determine the desired vertical exaggeration and scale the profile sheet appropriately. **Scaling the profile sheet** means drawing a series of equally spaced parallel lines the length of the trace. The distance between the lines is determined by the vertical exaggeration. For example, if the horizontal scale is 1 in = 150 ft and the vertical exaggeration is × 10, the parallel lines would be spaced one inch apart and each line would represent a change in depth of 15 ft.

    VE = Horizontal Scale /Vertical Scale
    10 = (1 in = 150 ft)/Vertical Scale
    Vertical Scale = (1 in = 150 ft)/10
    Vertical Scale = (1 in = 15 ft)

3.  Label the lowest line on the profile sheet with the value of one contour interval deeper than the deepest contour the trace intersects. For example, if the deepest contour crossed by the trace is −50 ft, label the lowest line −60, if the contour interval is 10 feet. This is done because part of the profile likely extends slightly below the deepest contour crossed. Sequentially number each line on the profiling sheet according to the contour interval. The top line should be one contour shallower than the shallowest contour crossed by the trace because part of the profile likely extends slightly above that depth. If too many closely spaced contours are present along the trace, only use index contours. Although this will not produce as accurate a picture, it will still present an adequate profile.

4.  Place the profile sheet so that the profile scale lines are parallel to the trace (Figure 2-11). Holding the sheet firmly in place, wherever a contour line intersects the trace, sketch a faint line perpendicularly upward from the trace to the scaling line on the profile sheet that corresponds to that depth.

5.  After all contours along the trace have been marked on the profile sheet, connect the ends of the perpendicular lines with a smoothly curving line. This line is the bathymetric profile along the trace.

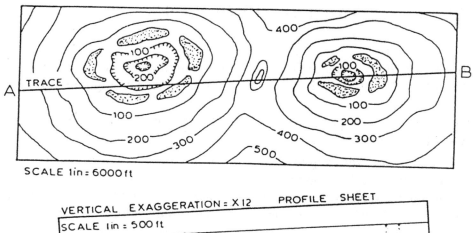

SCALE 1in = 6000ft

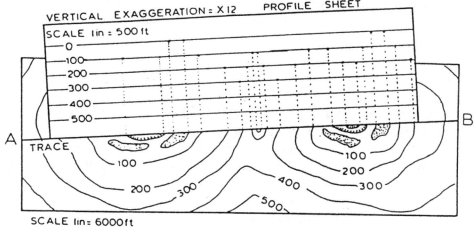

SCALE 1in = 6000ft

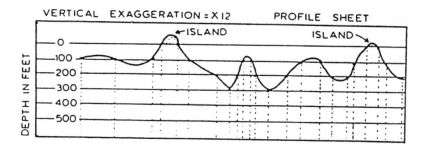

Figure 2-11. Sequence in producing a bathymetric profile. 1. Draw trace on chart. 2. Place profile sheet along trace. 3. Draw perpendicular lines from the intersection of each contour line and the trace to the appropriate depth on the profile sheet. 4. Draw a smoothly curving line connecting the ends of the perpendicular lines. This curving line is the vertically exaggerated profile of the sea floor along the trace.

## EXERCISE 2. BATHYMETRIC PROFILES

1. Sandy Harbor chart
    a. On the Sandy Harbor chart (Figure 2-8), contoured in Exercise 1, are the letters DD′ and EE′. The dashed lines connecting the corresponding pairs of letters are traces for profiles.
    b. Using the appropriate profile sheets provided (Figure 2-12), draw the profile between DD′ and EE′.
    c. Determine the vertical exaggeration for each.
        • To determine the horizontal verbal scale for the Sandy Harbor chart, use a centimeter scale (ruler) and measure along the length of the chart scale. From 0 to about 1.5 km on the chart scale is a distance of 2.0 cm on the centimeter scale. This provides the verbal scale of 2.0 cm = 1.5 km or 1 cm = 0.75 km.
        • To determine the vertical scale for the profiling sheet, use a centimeter scale (ruler) and measure vertically from the 0 fathom profile line downward to where another profile line intersects a division on the centimeter scale. For profile sheet DD′, the 3.5 fathom line intersects the 3.0 cm line. This means that the vertical verbal scale is 3 cm = 3.5 fathoms or 1 cm = 1.17 fathoms. Convert units as needed and substitute values into the formula to determine vertical exaggeration.

2. Portion of the southern Pacific Ocean
    a. On the southern Pacific Ocean chart (Figure 2-9) contoured in Exercise 1 are the letters FF′. Connect these letters with a straight line, the trace for the profile.
    b. Using the appropriate profile sheet provided (Figure 2-12), draw the profile between FF′.
    c. Determine the vertical exaggeration for each.

## EXERCISE 3. DISTORTION FROM VERTICAL EXAGGERATION

1. Using the Sandy Harbor chart (Figure 2-8), draw the profile along trace EE′ on the profile sheet provided on Figure 2-13 and determine the vertical exaggeration. Compare this profile to the profile of EE′ drawn on Figure 2-12 to answer the following questions.
    a. Explain how vertical exaggeration alters the shape of the profiles.
    b. If vertical exaggeration distorts slopes, why does it not distort flat areas?
    c. Would vertical exaggeration distort vertical cliffs? How? Why?

**D**  Depths in fathoms          Vertical Exaggeration = _____                                    **D′**

- 0
-
- 1
-
- 2
-
- 3
-
- 4
-
- 5

**E**  Depths in fathoms          Vertical Exaggeration = _____                                    **E′**

- 0
- 1
- 2
- 3
- 4
- 5
- 6
- 7
- 8
- 9

**F**  Depth in meters            Vertical Exaggeration = _____                                    **F′**

- 0
- 100
- 200
- 300
- 400
- 500
- 600
- 700
- 800
- 900
- 1000
- 1100
- 1200
- 1300
- 1400
- 1500
- 1600
- 1700
- 1800
- 1900

Figure 2-12.  Profile sheets for Exercise 2 for the construction of profiles DD′, EE′ and FF′.

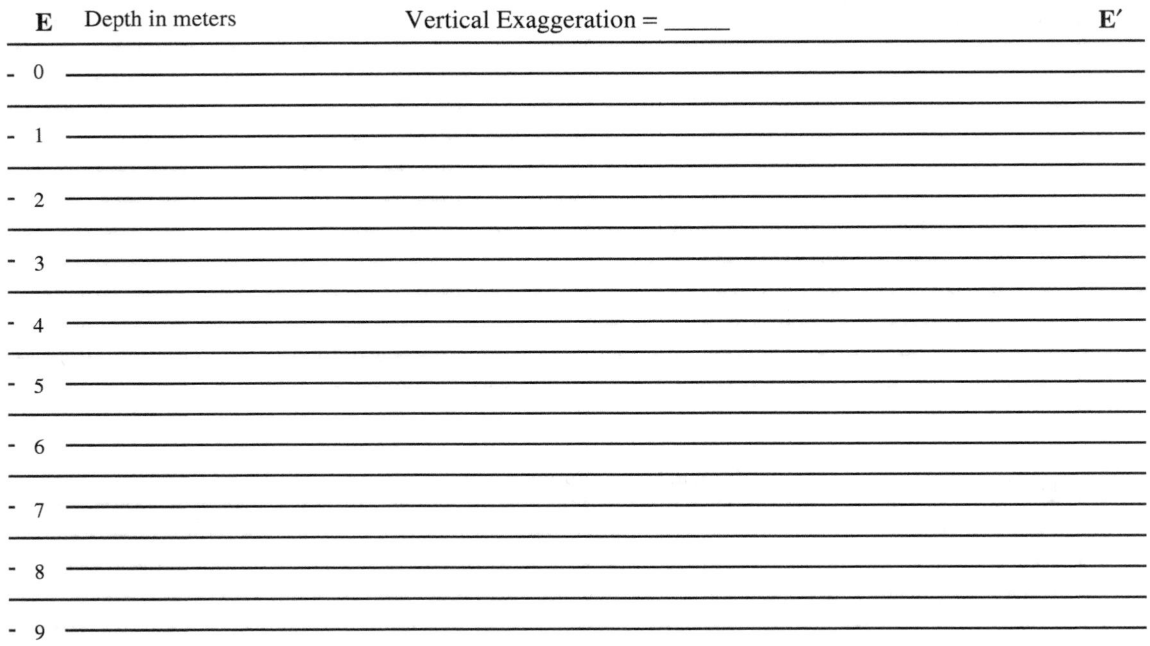

| E | Depth in meters | Vertical Exaggeration = _____ | E′ |

Figure 2-13.  Profile sheet for comparing effect of distortion from vertical exaggeration.


## 2-7.  SCIENTIFIC NOTATION

Scientific notation is a short-hand method of writing extremely large or small numbers. Mathematically, it consists of multiplying a number by itself a certain number of times. For example, $x^n$ (read x to the **n power** and n is referred to as the **exponent** of x) means that the number x is multiplied by x, n times. If $x = 2$ and $n = 4$, then:

$$2^4 \text{ is } 2 \times 2 \times 2 \times 2 = 16 \text{ or } 2^4 = 16.$$

If the value of n is a negative number, then it is 1/x multiplied by itself n times. If $x = 2$ and $n = -4$, then:

$$2^{-4} \text{ is } 1/2 \times 1/2 \times 1/2 \times 1/2 = 1/16 \text{ or } 2^{-4} = 1/16$$

In scientific notation, we are most commonly interested in eliminating the zeros present in large or small numbers. For each position a decimal point is moved to the left, it is as though the number has been divided by 10. To retain the same value, the number must be multiplied by 10. Thus, 2000 is the same as $200 \times 10$ or $20 \times 10 \times 10$ or $2 \times 10 \times 10 \times 10$. Because $10 \times 10 \times 10$ is the same as $10^3$, 2000 is the same as $2 \times 10^3$.

For each position a decimal point is moved to the right, it is as though the number has been multiplied by 10. To retain the same value we must multiply the number by 1/10. Thus, 2 is the same as $20 \times 1/10$ or $200 \times 1/10 \times 1/10$ or $2000 \times 1/10 \times 1/10 \times 1/10$. Because $1/10 \times 1/10 \times 1/10$ is the same as $10^{-3}$, 2 is the same as $2000 \times 10^{-3}$.

Large numbers such as 12,000,000,000,000 can be written as $1.2 \times 10^{13}$. Extremely small numbers such as 0.00000000234 can be written as $2.34 \times 10^{-9}$. Typically, the decimal is placed behind the first digit on the left. For numbers such as 10, 100, 1000, etc., it is not necessary to write $1 \times 10^{n}$, but only $10^{n}$. The "1 x" is understood.

## EXERCISE 4.  SCIENTIFIC NOTATION

1. Convert the following numbers into scientific notation.
   a. Earth's average distance from the Sun, 92,600,000 mi  _____
   b. Earth's diameter, 8000 mi  _____
   c. Earth's surface area, 254,000,000 mi$^2$  _____
   d. Ocean depth, 11,000 m  _____
   e. Volume of 1 gram = 0.035 oz  _____
   f. Wavelength of gamma rays, 0.00000000003 m  _____
   g. Population of North America, 397,000,000  _____
   h. Mass of an electron relative to a proton, 0.001  _____
   i.  Mass of an electron, 0.00000000000000000000000000091067 g  _____
   j. Total weight of the atmosphere, 57,000,000,000,000,000 tons  _____

2. Convert the following from scientific notation into a regular number.
   a. Mass of the Earth, $5.983 \times 10^{24}$ kg  _____
   b. Velocity of light, $2.99776 \times 10^{10}$ cm/sec  _____
   c. Unit mass of an atom, $1.66035 \times 10^{-24}$ g  _____
   d. Volume of water in the ocean, $1.4 \times 10^{9}$ km$^3$  _____
   e. Size of clay particles, $1 \times 10^{-4}$ mm  _____

## MATHEMATICAL FUNCTIONS

### Addition and Subtraction

When adding or subtracting numbers written in scientific notation, it is necessary for the two numbers be to the same power (have the same exponent). To accomplish this, move the decimal point as necessary and change the exponent to correspond.

Example 1. $1.45 \times 10^5 + 2.67 \times 10^4 = ?$
   Change $1.45 \times 10^5$ to $14.5 \times 10^4$ or change $2.67 \times 10^4$ to $0.267 \times 10^5$ and then add the two numbers.

   $14.5 \times 10^4 + 2.67 \times 10^4 = 17.17 \times 10^4$ or $1.45 \times 10^5 + 0.267 \times 10^5 = 1.717 \times 10^5$
   The two answers are equal.

Example 2. $2.13 \times 10^2 + 400 \times 10^{-2} = ?$

  Change $2.13 \times 10^2$ to $21,300 \times 10^{-2}$ or change $400 \times 10^{-2}$ to $0.04 \times 10^2$ and then add the two numbers.

  $21300 \times 10^{-2} + 400 \times 10^{-2} = 21700 \times 10^{-2}$   or   $2.13 \times 10^2 + 0.04 \times 10^2 = 2.17 \times 10^2$

  The two answers are equal.

### *Multiplying or Dividing*

When multiplying numbers written in scientific notation, add the exponents together. When dividing numbers written in scientific notation, subtract the exponent of the denominator from the numerator.

Examples: $2 \times 10^2 \times 3 \times 10^3$ is the same as $200 \times 3000$, which equals $600,000$ or $6 \times 10^5$

  $(2 \times 10^8) \times (3 \times 10^4) = 6 \times 10^{12}$       (exponent $8 + 4 = 12$)
  $(2 \times 10^8) \times (3 \times 10^{-6}) = 6 \times 10^2$       (exponent $8 + -6 = 2$)
  $(2 \times 10^{-8}) \times (3 \times 10^{-6}) = 6 \times 10^{-14}$       (exponent $-8 + -6 = -12$)

Examples: $6 \times 10^6 \div 2 \times 10^2$ is the same as $6,000,000/200$ and equals $30,000$ or $3 \times 10^4$

  $(4 \times 10^8)/(2 \times 10^4) = 2 \times 10^4$       (exponent $8 - 4 = 4$)
  $(4 \times 10^8)/(2 \times 10^{-6}) = 2 \times 10^{14}$       (exponent $8 - -6 = 14$)

## EXERCISE 5.  CALCULATIONS IN SCIENTIFIC NOTATION

1. Mount Everest is about $2.9 \times 10^4$ ft above sea level. The deepest trench is $34.2 \times 10^3$ ft below sea level. If Mount Everest was dropped into the trench, how deep would the water be above it?   _____

2. The average distance from the Earth to the Moon is $2.38 \times 10^6$ mi. The average distance from the Earth to the Sun is $0.926 \times 10^8$ mi. What is the distance between the Moon and the Sun when the Moon is (a) between the Earth and the Sun, and when the Earth is (b) between the Moon and the Sun?
   a. _____   b. _____

3. The velocity of light is $2.99776 \times 10^{10}$ m/sec. (a) How far will light have traveled in 100 seconds?  (b) In a year?   a. _____   b. _____

4. The simplest hydrogen atom consists of only a proton and electron. If the mass of this simple hydrogen atom is $1.66035 \times 10^{-24}$ g, and the mass of an electron is $9.1066 \times 10^{-28}$ g, what is the mass of the proton?   _____

5. The population of the Earth is about $6 \times 10^9$, the land area of the Earth is about $148.847 \times 10^6$ km$^2$. How much land is there for each person?   _____

# Laboratory 3

# Ocean Basin Physiography and Plate Tectonics

## 3-1. THE OCEAN BASIN

The **ocean basin** is the vast depression in the Earth's surface bordered on all sides by continents and filled with a continuous body of salt water, the ocean. Present-day oceans have overflowed the basin and flooded the edge of the continents to a depth of about 200 m. The ocean basin and associated submerged continental edges are not flat. They display numerous features, varying from submerged mountain chains to deep trenches that extend several kilometers below the adjacent sea floor.

A sequence of major bathymetric (sea floor) regions may be identified that extends seaward from the coast. These include the continental shelf, continental slope, continental rise, abyssal plains, abyssal hills, ocean ridge crest, and ocean trenches (Figure 3-1). Additionally, numerous minor features, such as submarine canyons, submarine fans, knolls, and seamounts, are recognized within some of these regions.

The **continental shelf** is the flooded edge of the continent (Figure 3-1). It extends from the beach to a depth between 130 and 200 m. Where the shelf is wide, the sea floor generally slopes at 1° to 2°, but where the shelf is narrow, the sea floor descends more steeply. The shelf usually has a gently undulating surface and may exhibit many features, such as hills and valleys, as seen on the adjacent exposed coastal area.

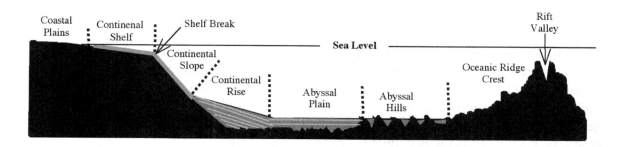

Figure 3-1. Cross section showing bathymetric features of the sea floor. Gray areas are sediments.

The **continental slope** (Figure 3-1) is the inclined margin of the continent and extends from the **shelf break** (seaward edge of the shelf) to the continental rise at a depth of about 2–3 km. The slope is the boundary between the continental landmass and the ocean basin. As with the shelf, the slope may be broad and descend at a gentle angle of 4° to 6° or, if adjacent to a trench, narrow and descend more precipitously.

The **continental rise** (Figure 3-1) forms a transition between the continental slope and abyssal plain. Produced by the accumulation of mainly land-derived sediment at the base of the slope, the rise obscures the contact between continent and ocean floor. These sedimentary accumulations may be up to several kilometers thick, tens of kilometers wide and incline between 1° and 5°.

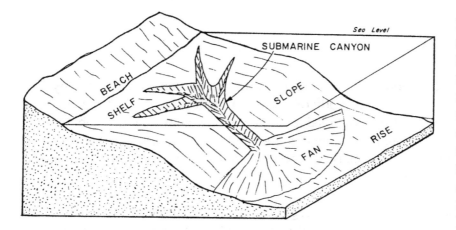

Figure 3-2. Submarine canyon and fan.

Locally, deep **submarine canyons** have been eroded into the shelf, slope and rise (Figure 3-2) by **turbidity currents**, density currents of sediment-laden water that flow rapidly downslope under the force of gravity. Deposition of sediments at the mouth of the canyons may form enormous fan-shaped accumulations called **submarine fans** (Figure 3-2). Submarine fans form part of the continental rise.

Seaward from the continental rise is the **abyssal plain** (Figure 3-1), a relatively flat featureless area where sediments have buried most sea floor irregularities. Isolated hills may protrude through the sedimentary cover. They are called **knolls** if less than 1000 m high or **seamounts** if higher than 1000 m. The number of knolls and seamounts increases seaward. They become the dominant feature in the **abyssal hills** (Figure 3-1) because sedimentary cover is too thin to completely bury the flank of the oceanic ridge.

**Oceanic ridges** (Figure 3-1) are volcanic in origin. They form mountain chains that extend the length of the ocean basins. Although the ridges extend from continent to continent and form most of the ocean floor, the outer edges usually are buried under the sediments of the continental rise, abyssal hills, and abyssal plains. The highest peaks generally are near the axis of the ridge and form an area called the **ridge crest**. Extending along the ridge crest is the narrow **rift valley** (Figure 3-1).

**Transform faults** (Figure 3-3) are located between the offsets in the oceanic ridge rift valleys. Along these faults, rocks are under extreme stress, and periodically break and slide past each other. **Fracture zones** are bands of distorted and broken rocks that extend outward from the transform faults. Originally they were part of the transform faults but have now progressed to where adjacent parts of the seafloor are moving in the same direction. Transform faults together with the fracture zones appear as gashes or scars extending

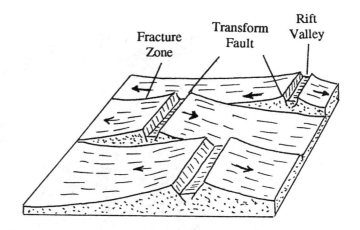

Figure 3-3. Transform faults are only between ridge segments where plates are moving in opposite directions.

across the sea floor, except where buried by sediment. They are generally aligned perpendicularly to the oceanic ridge rift valleys.

**Oceanic trenches** are very deep, long, narrow, relatively steep-sided depressions in the ocean basin. They usually occur adjacent to the continents, but may form well away from the continental masses.

## 3-2. PHYSIOGRAPHIC MAPS

**Physiographic maps** attempt to provide a three-dimensional perspective view of a region. Variation in elevation is indicated by differences in shading. Because perspective changes with distance from the point of observation, shading will also vary. For maps used in these exercises, the point of observation is always from the bottom edge of the page.

## EXERCISE 1. PHYSIOGRAPHIC MAP

Features of the Atlantic Ocean sea floor (Figure 3-4).

1. Label and shade the following major land masses: North America, South America, the Greater Antilles (Cuba, Hispaniola and Puerto Rico) and the Lesser Antilles.

2. Identify the following features
    A. _____ (flooded edge of the continent)
    B. _____ (inclination toward the sea floor)
    C. _____ (thick sedimentary accumulation at the continent's base)
    D. _____ (featureless part of the sea floor)

Figure 3-4.  Physiographic map of a portion of the Atlantic Ocean Basin.

E. _____ (low hills, exposed tops of volcanoes)
F. _____ (tall volcanic cones higher than 1000 m)
G. _____ (submerged, deep valley)
H. _____ (local, massive accumulation of sediment at canyon mouth)
I. _____ (deep, long depression in sea floor)
J. _____ (mountainous central region of sea floor)
K. _____ (valley in center of mountainous region)
L. _____ (active offset of mountainous region)
M. _____ (inactive offset of mountainous region)

## 3-3. PLATE TECTONICS

Heat generated within the Earth migrates outward toward the surface. In the semi-solid asthenosphere, the escaping heat establishes a series of convection cells. Friction between the flowing rocks of the asthenosphere and the base of the lithosphere creates enormous stress on the overlying **lithospheric plates** (Figure 3-5). When sufficient force accumulates, the plates are moved by the convection currents. In areas where heated asthenosphere rises and flows outward in opposite directions, adjacent plates are under **tension** and move apart forming a **divergent plate edge**. In areas where cooled asthenosphere subsides, adjacent plates are **compressed** and one plate will slide below the other, forming a **convergent plate edge**. The movement of lithospheric plates is called **plate tectonics**.

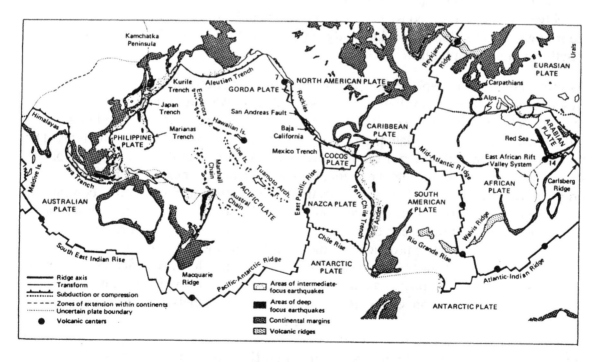

Figure 3-5. Lithospheric plates of the Earth. Note plate edges.

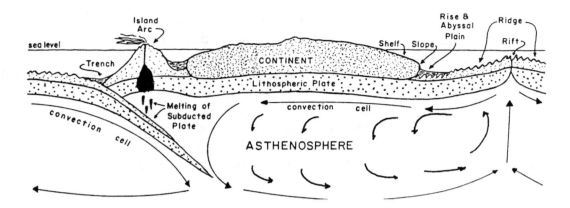

Figure 3-6.  Cross section of the outer portion of the Earth showing convection in the asthenosphere.

Ocean basin features result primarily from plate tectonics and sedimentation. Plate tectonics produce sharp, jagged features, but sedimentation tends to either bury them or smooth their outlines. Where ascending heat produces two diverging convection cells, the lithosphere tilts outward and the ocean floor bows upward (Figure 3-6). The asthenosphere in these convection cells flows away from the central area and they slowly drag the overlying plates in opposite directions. As the plates split apart, a process called rifting, a **rift valley** (Figure 3-6) forms. Periodically, lava is injected into the valley forming new rocks, which are added to the plates' edges as the ocean basin expands. The injected lava can form undersea volcanoes (seamounts) as part of the oceanic ridge system, or massive lava flows. Thus, the oceanic ridge system is bathymetrically high at the ridge crest because it consists of a series of volcanic mountains and because the ocean floor bows upward along the ridge. Ridge systems and rift valleys are offset along their lengths by transform faults.

As rocks in the asthenosphere flow along the base of the lithosphere, they cool and slowly descend. The ocean floor also tends to subside away from the ridges. In addition to being bathymetrically lower than the ridge crest, these areas are older and more sediment will have accumulated, especially adjacent to the continental landmasses. The gradual change in bathymetry from oceanic ridge crest to abyssal hills to abyssal plains and continental rise represents the gradual subsidence of the ocean floor and the progressive burial of ridge features by accumulating sediment. Some young continental slopes represent the modified surface along which the continental landmasses split as rifting began to form the ocean basin (e.g., South America—Africa; North America—Africa). Other slopes were produced by the complex interaction of oceanic and continental plate movement (e.g., west coast of South and North America). Subsidence of the rocks in the asthenosphere can also bend the edges of the continents downward causing them to flood more easily, forming the continental shelves (Figure 3-6).

In addition to the major convection cells associated with the oceanic ridge systems, local super-heated fountains of mantle rock, called **mantle plumes**, occur within the asthenosphere (Figure 3-7, top). The geographic location above a mantle plume is called

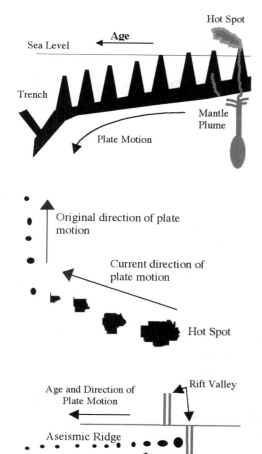

Figure 3-7. Top: Cross section showing mantle plume and sequence of volcanoes produced. Middle: Map view of chain of volcanoes produced at hot spot, indicating change in plate motion. Bottom: Map view showing an aseismic ridge trending away from a rift valley.

a "**hot spot**" and its presence is usually indicated by greater than normal volcanic activity. Typically, large volcanoes gradually form at the hot spot. Although the location of the plume and hot spot are believed to remain relatively stationary over time, the lithospheric plate gradually moves across the plume and a linear series of seamounts and volcanic islands may be generated. The youngest and most active volcano is the one above the plume. The volcanoes become progressively older and less active away from the hot spot. By plotting the volcanoes produced by a hot spot, it is possible to determine the direction in which the plate is moving and has moved in the past (Figure 3-7, middle). A chain of larger than normal volcanoes produced by a mantle plume near a ridge crest is called an **aseismic ridge** (Figure 3-7, bottom). Aseismic ridges tend to be aligned perpendicularly to the ridge crest and parallel to transform faults and fracture zones. Aseismic ridges can also indicate if there has been a change in direction of plate motion.

Where plates converge, the sea floor is greatly depressed and forms a trench (Figure 3-6). Eventually, one plate will be pushed beneath (**subducted**) the other. The plate that descends into the Earth is called the **subducted plate** and the plate below which it sinks is the **overriding plate**. Areas of trench formation are called **subduction zones**. Because continental rock is less dense and much thicker than oceanic crust, continents always override ocean floor in a subduction zone.

As the subducted plate slides deeper into the asthenosphere, it melts. Some of the molten rock migrates upward because it is less dense than the rocks of the asthenosphere. As the molten mass melts its way through the overlying lithosphere, it forms a series of volcanoes. If the overriding plate in the area of subduction is ocean floor, the volcanoes form a chain of islands, called an **island arc**, which parallels the trench. If the overriding plate in the area of subduction is a continent, a volcanic mountain chain paralleling the trench forms on the continent. Subduction can eventually eliminate an entire ocean basin causing continents to collide, forming large, structural, nonvolcanic

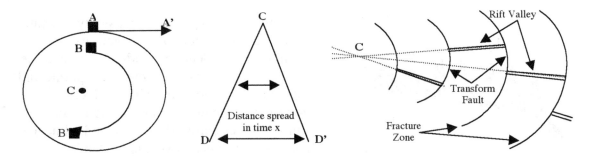

Figure 3-8. (Left) A–A, object moving in a straight line would leave the surface of the Earth. B–B', object moving across Earth's surface travels in an arc with C as the center or spreading pole. (Middle) D–D' with greater distance from C, the linear rate of spreading increases. (Right) Lines drawn through each rift valley segment converge at location of spreading pole for the plate.

mountain chains. The collision of India and Asia formed the Himalayas and eliminated the ocean basin between them.

Because the Earth is spherical in shape, no object can move in a straight line without leaving the Earth's surface (Figure 3-8, left). Any object, including a lithospheric plate, moving across the Earth's surface travels in an arc, a curved line that is part of a circle.

All circles have a central point about which an object rotates as it travels in the circle. The point on the Earth's surface about which a plate revolves as it moves in an arc is called the **spreading pole** Each plate has its own spreading pole and if direction of plate movement changes, the location of the spreading pole also changes. In general, the farther an area is from the spreading pole, the faster the rifting up to 90° away from the spreading pole (Figure 3-8 center). A spreading pole can be located using the rift valleys. Rift valley segments point toward the spreading pole of the plate. Lines drawn through each rift valley's segment should converge at the location of the spreading pole for the plate (Figure 3-8 right).

In summary:

1. Edges of lithospheric plates are located at rift valleys, transform faults, trenches, and structural mountain chains.
2. Lithospheric plates are slowly dragged forward by fluid currents within the asthenosphere. Plates move *away from the ridges, toward the trenches, and parallel to the transform faults and the recent segments of volcanic chains originating from hot spots.*
3. New rock is added to the edges of the plates at the ridge crest. Sea floor rocks at the crest of the oceanic ridge are therefore the youngest and the sea floor is progressively older away from the ridge.
4. Because the sea floor is older near the continents, sediments have had a longer time to collect and bury ridge features in these areas.

5. Sea floor is destroyed in ocean trenches. The overriding plate will contain either an island arc or volcanic mountain chain.
6. Subduction may eventually eliminate an ocean basin, allowing continents to collide and forming high, non-volcanic, structural mountain chains.

## EXERCISE 2.  PLATE TECTONICS

1. Physiographic map of a portion of the Atlantic Ocean (Figure 3-4).
   a. On the map, draw lines indicating the edges of the plates.
   b. Place arrows on the map indicating the direction in which the plates are moving.
   c. For the portion of the North Atlantic displayed on the map, where would the oldest part of the sea floor be located?  _____
   d. Why is the ridge in the middle of the ocean in the North Atlantic (see Figure 3-5)?



   e. Where would the following be expected to occur in this ocean basin?
      1. Oldest sediment  _____
      2. Thickest sediment _____
      3. Volcanic activity  _____
   f. A linear series of seamounts along line N were produced by a mantle plume as the plate slowly moved across a hot spot. Where would the oldest volcano in the series be located?   Why?



   g. Note the trench just to the west of the Antilles. Describe the plate motion in this area. Which plate is beings subducted? Give evidence for your answer.



2. Physiographic map of a portion of the Pacific Ocean (Figure 3-9).
   a. Label and shade the parts of Asia and Australia shown on the map.
   b. Label the following on the map and outline the plate edges (see Figure 3-5).
      1. Pacific Plate          5. Philippine Trench
      2. Philippine Plate       6. Kurile Trench
      3. Australian Plate       7. Japan Trench
      4. Eurasian Plate         8. Marianas Trench
   c. Label some of the areas that are abyssal plains.
      1. Why do abyssal plains appear to be absent when a trench is adjacent to a continent?

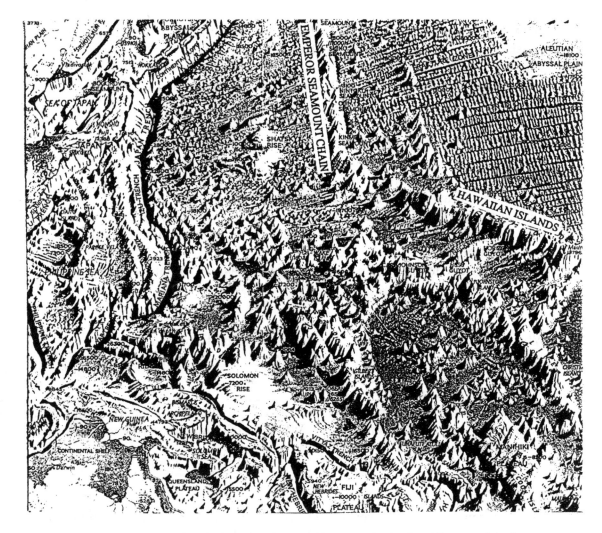

Figure 3-9. Physiographic Map of a portion of the Pacific Ocean basin. South is toward the bottom.

2. Why are abyssal plains absent in much of the northwestern Pacific Basin?

d. Note the Emperor Seamounts and Hawaiian Island chain. Only the southeastern-most islands of Hawaii presently have active volcanoes.

1. What would produce a series of volcanic islands in the central portion of the Pacific Plate?_____

2. Where would the oldest seamount be in the Emperor Chain?

3. If all of these seamounts were produced by the same phenomenon, in what direction do they indicate the plate is currently moving? _____

4. Is there evidence that the plate once moved in a different direction? If so, in what direction did it previously move? _____

5. The northwest end of the Emperor Seamounts is in subpolar waters. Yet, samples of deep sediment contain fossil shells of organisms typical of a more tropical environment. How can this be explained by plate tectonics?

   e. Using a world map and Figure 3-5, identify several of the island arcs present on the map and the trenches with which they are associated.

   f. Is the Eurasian or the Pacific plate being subducted along the western edge of the ocean? Give evidence to support your answer.

   g. Compare and contrast the Atlantic and Pacific Ocean basins in terms of the following:
      1. Location of oceanic ridges.
      2. Location of trenches.
      3. Size of the abyssal plains.
      4. Size of the continental rises.

3. Although physiographic maps provide a unique view of the ocean floor, one rarely has such detailed information. More commonly, one must work from simplified line maps in interpreting plate tectonics. The questions below refer to the following line map. See legend below map. Circle the letter of each true statement.
   a. Yehling Islands are an aseismic ridge.
   b. Features K are transform faults.
   c. The area labeled D is the continental slope.
   d. Pershing Islands along the northern coast of Continent B are an island arc.
   e. Lack Islands to the southeast of Continent A are an island arc.
   f. Island G should be older than Island N in the Brock Island Chain.
   g. Feature L are transform faults.
   h. Island Z is older than X in the Yehling Island Chain and that is why it is larger.
   i. Evidence suggests that Continent A is being rifted.
   j. Barron Island chain originate at a hot spot.
   k. Island M is older than Y in the Barron Island Chain.
   l. Ty Lakes on Continent A most likely are located in a rift valley.
   m. Sati Mountains on Continent A are most likely volcanic in origin.
   n. P is a transform fault.
   o. Coonan Mountains on Continent B are most likely volcanic in origin.
   p. Toyke Islands and Brock Islands are on the same plate.
   q. Plate 1 once moved to the northeast but now moves mainly northward.
   r. Plate 4 is mainly moving southward and converging with plate 3.
   s. Island W is probably much older than Island C in the Pershing Chain.
   t. **On the map**, locate the spreading pole for plate 2.
   u. Draw arrows on the map indicating the direction in which the plates are moving.

North

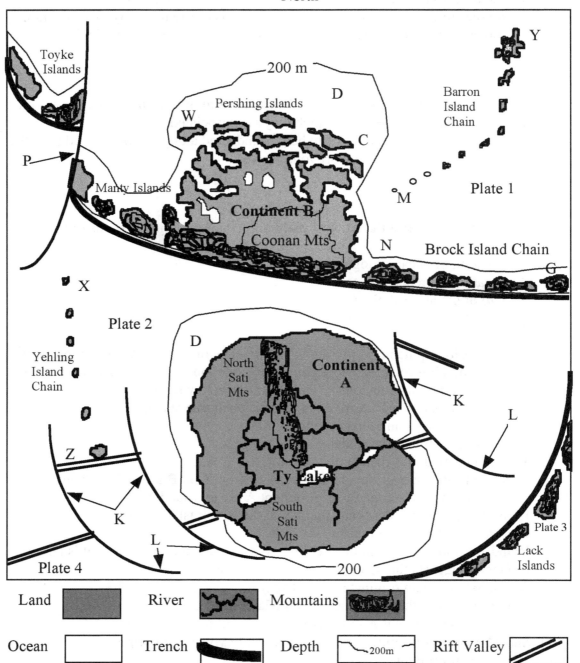

Land    River    Mountains

Ocean    Trench    Depth 200m    Rift Valley

## 3-4.  RATE OF OCEAN BASIN EXPANSION

It is possible to determine the average rate at which the sea floor is spreading outward from the rift valley on the oceanic ridge if one can date the volcanic rocks of the sea floor and measure their distance from the crest of the oceanic ridge. Average rate of sea floor spreading is equal to the distance the rocks have traveled from the rift valley divided by the time it has taken them to reach the point where they are located.

Average Rate of Sea Floor Spreading = Distance/Time

Distance/time will indicate the average rate of spreading for one side of the oceanic ridge. If the ocean basin is symmetrical, as is the Atlantic Ocean, spreading will be occurring on both flanks of the ridge at the same relative average rate. Average rate of ocean basin expansion will be equal to twice the average rate of sea floor spreading.

Average Rate of Ocean Basin Expansion = 2(Distance/Time)

Example 1: Volcanic sea floor rocks are dated as being 20 million years old. They are located 400 km from the rift valley in the ridge crest. What are the average yearly sea floor spreading rate and the average yearly ocean basin expansion rate over the last 20 million years for this ocean basin?

$$\text{Average Rate of Sea Floor Spreading} = \text{Distance/Time}$$
$$= 400 \text{ km}/20{,}000{,}000$$
$$= 400 \text{ km}/20 \times 10^6 \text{ yr}$$
$$= 20 \times 10^{-6} \text{ km/yr}$$
$$= 2 \times 10^{-5} \text{ km/yr}$$
$$= 2 \times 10^{-5} \text{ km/yr} \times 10^3 \text{ m/km} \times 10^2 \text{ cm/m} = 2 \text{ cm/yr}$$
$$\text{Average Rate of Ocean Basin Expansion} = 2(\text{Distance/Time})$$
$$= 2(2 \text{ cm/yr})$$
$$= 4 \text{ cm/yr}$$

Sea floor spreading rates and ocean basin expansion rates may increase or decrease over time. Sea floor spreading may even cease completely for a while and then restart. This is obscured by the average spreading rates because all of the various rates within an interval are represented by a single number, the average. The farther a sea floor rock is from the rift valley and the older the rock, the more the average spreading rate obscures the true spreading rate history. To determine the true spreading rates for an ocean basin, it is necessary to work with segments of the sea floor and determine the average spreading rate between each segment.

Example 2: Volcanic sea floor rocks from three locations arranged outward from the rift valley have been dated and their distance to the crest measured. Determine the average spreading rate between each location.

| Location | Age (million years) | Distance to Crest (kilometers) |
|----------|---------------------|--------------------------------|
| A | 15 | 525 |
| B | 32 | 1392 |
| C | 40 | 1488 |

The spreading rate between location A and the rift valley is determined the same as in the first example.

Average Rate of Sea Floor Spreading = Distance/Time
$$= 525 \text{ km}/15,000,000$$
$$= 525 \text{ km}/15 \times 10^6 \text{yr}$$
$$= 35 \times 10\text{-}6 \text{ km/yr}$$
$$= 35 \times 10\text{-}6 \text{ km/yr} \times 10^3 \text{ m/km} \times 10^2 \text{ cm/m} = 3.5 \text{ m/yr}$$

To determine the spreading rate between location A and B, first determine the distance between them and the time interval over which this spreading has occurred. Then solve the problem in the same fashion as the example above.

Spreading Distance = B's distance from rift − A's distance from rift
Spreading Distance = 1392 km − 525 km
Spreading Distance = 867 km

Time interval of spreading = Age of B − Age of A
Time interval of spreading = $32 \times 10^6 \text{yr} - 15 \times 10^6 \text{yr}$
Time interval of spreading = $17 \times 10^6 \text{yr}$

Average Sea Floor Spreading Rate between A and B = $867 \text{ km}/17 \times 10^6 \text{yr}$

$$= 51 \times 10^{-6} \text{ km/yr}$$
$$= 51 \times 10^{-6} \text{ km/yr} \times 10^3 \text{ m/km} \times 10^2 \text{ cm/m}$$
$$= 5.1 \text{ cm/yr}$$

C is solved in the same fashion. Average sea floor spreading rate between B and C is 1.2 cm/yr. Thus, the average sea floor spreading rate has varied between 1.2–5.1 cm/yr.

# EXERCISE 3.  THE RATE OF SEA FLOOR EXPANSION

Volcanic sea floor rocks from six locations arranged outward from the rift valley have been dated and their distance to the rift valley measured.

| Location | Age (million years) | Distance to Rift (kilometers) |
|---|---|---|
| A | 10 | 100 |
| B | 15 | 375 |
| C | 33 | 807 |
| D | 39 | 1233 |
| E | 43 | 1353 |
| F | 70 | 1674 |

1. What is the average rate of sea Floor spreading for this entire section of ocean basin?

2. What is the average rate of ocean basin expansion for the entire basin?

3. Determine the average rate of sea floor spreading and the average rate of ocean basin expansion for each segment of the ocean basin.

| Segment | Average Sea Floor Spreading Rate (cm/yr) | Average Basin Expansion Rate (cm/yr) |
|---|---|---|
| A–crest | _____ | _____ |
| A–B | _____ | _____ |
| B–C | _____ | _____ |
| C–D | _____ | _____ |
| D–E | _____ | _____ |
| E–F | _____ | _____ |

4. How representative is the average rate of sea floor spreading compared to the true history of basin expansion?

# Laboratory 4

# Bottom Sediment Charts

## 4-1. BOTTOM SEDIMENT CHARTS

**Bottom sediment charts** show the distribution and type of material on the sea floor. This information is important in determining the geology of an area, the suitability of the bottom for anchorage of ships, bridge supports, oil platforms, cables, etc., and can even be useful in determining one's location. Data on bottom sediments in deep parts of the ocean can only be obtained by lowering devices that collect samples. This is both slow and costly. Thus, bottom sediment charts are usually developed from much less data than are bathymetric charts.

Proper interpretation of bottom samples requires knowledge of factors controlling sediment distribution. For water- and wind-transported sediment, particle size is a major control. Erosion of sediment by wind or water results when fluid velocity produces sufficient turbulence to overcome **inertia** (resistance to being moved). As velocity increases, turbulence increases. Once sediment is in motion, continued transportation requires only sufficient turbulence to overcome a grain's **settling velocity** (how rapidly the grain sinks). Deposition occurs when turbulence is less than settling velocity and the particle sinks to the bottom.

## 4-2. HJULSTROM DIAGRAM

The relationship of particle size and fluid velocity to erosion, deposition, and transport is summarized in the **Hjulstrom Diagram** (Figure 4-1). On this diagram, particle size increases to the right and fluid velocity increases toward the top. The broad U-shaped line separating the areas labeled Erosion and Transportation is the *minimum velocity* for erosion. For a given particle size, any velocity on or above this line is strong enough to cause erosion. The sloping line separating areas labeled Deposition and Transportation represents the *maximum velocity* at which deposition can occur. For a given particle size, a velocity on or below this line is too slow to keep a particle in transport. Deposition then occurs. The Hjulstrom diagram does not address the dissolution of grains by water.

From this diagram, the following can be concluded about particles of the same density:

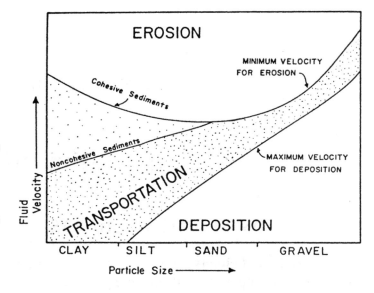

Figure 4-1.  Hjulstrom Diagram.

1. Because of inertia, a greater fluid velocity and turbulence are needed to erode a particle than to transport it, regardless of particle size.
2. For sand and larger particles, the fluid velocity and turbulence necessary to erode and transport increases with particle size. This is because larger particles weigh more and settle faster.
3. For particles smaller than sand, the fluid velocity and turbulence needed to erode them increases as particle size decreases. This is because small particles tend to be **cohesive** (stick together). A greater velocity is needed to pull the grains apart. If the particles are noncohesive, velocity needed to erode decreases with size.
4. As fluid velocity and turbulence decrease, particles are deposited according to size with the largest first and smallest last. Because of weight, larger particles settle faster than small ones.

Numerous examples of these four principles operating in nature can be cited. Where a river flows into the ocean, its velocity gradually decreases. Larger particles are deposited near the river's mouth and finer material is carried far offshore before it settles to the bottom. Waves breaking on a shore will produce considerable turbulence. Larger particles will be moved about, but remain near the shore. Smaller particles will be unable to settle because the turbulence is greater than their settling velocity. These small particles will be transported to quieter areas in deeper water or into protected bays where they will be deposited. As turbidity currents flow from submarine canyons onto the continental rise and abyssal plain, current velocity gradually decreases. **Turbidites**, the sediments deposited by a turbidity current, grade from coarser material near the continental slope to finer sediments on the abyssal plains. They also show a gradation from coarser material at the bottom of the deposit to finer material at the top. In general, for water- and wind-transported sediments, grain size decreases away from the source area.

## 4-3. ORIGIN OF SEDIMENTS

Sediments can be divided into five groups based on their origin (Figure 4-2). **Terrigenous sediments** are produced by the weathering and erosion of rocks on land. These sediments tend to be concentrated near their source and decrease in size with distance. **Biogenic sediments** are mineralized structures (shells, teeth, etc.) produced by organisms. **Authigenic sediments**, also called **hydrogenic**, are particles produced by chemical reactions in sea water. **Evaporites** (salts) and **ferromanganese** and **phosphorite nodules** are the most common authigenic sediments in the ocean. **Volcanogenic sediments** are particles, not lava flows, ejected from volcanoes into the atmosphere, which then fall back to Earth. As with the terrigenous sediments, volcanogenic sediments tend to be concentrated near their source and to decrease in size with distance. **Cosmogenous sediments** consist mainly of the tiny grains that fall to Earth from space. Unlike other types of sediments, they potentially fall everywhere, but are seen in abundance only when other sediments are absent. These five types of sediments usually occur mixed together, but typically one type may dominate in a given deposit.

Normally, terrigenous or volcanogenic sediments dominate a near shore marine deposit. If the environment restricts the abundance of these sediments, biogenic sediment will be the most abundant. In deep oceanic environments where biogenic sediments are chemically unstable and dissolve or where little biogenic sediment is produced, hydrogenic sediments mixed with cosmogenic and wind-blown dust becomes common.

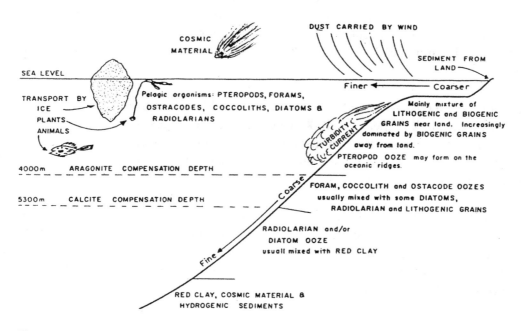

Figure 4-2.  Origin of sediments and distribution of biogenic sediments.

## 4-4.  BIOGENIC SEDIMENT DISTRIBUTION

Organisms generate a wide variety of biogenic sediment in the ocean, ranging in size from massive coral reefs hundreds of kilometers long to microscope particles. **Oozes** are the main biogenic sediment accumulating in the deep ocean basins. An ooze is a fine-grained sediment containing more than 30% biogenic material, primarily microscopic **pelagic** (floating) organisms. Distribution of biogenic sediments is controlled by the chemical stability of the mineral composing the grains, the environmental tolerance of the organisms' and the organism's abundance. Oozes typically occur far from land, where inflow of terrigenous sediment is greatly reduced. They are named for the dominant type of biogenic grain. The six types of common oozes are:

| Organism | Ooze | Mineral of Shell | Chemical Formula |
|---|---|---|---|
| Pteropod (pelagic snail) | Pteropod ooze | Aragonite | $CaCO_3$ |
| Coccoliths (algae) | Coccolith ooze | Calcite | $CaCO_3$ |
| Ostracod (shrimp-like) | Ostracod ooze | Calcite | $CaCO_3$ |
| Foraminifera (single cell animal) | Foraminiferal ooze | Calcite | $CaCO_3$ |
| Diatom (algae) | Diatom ooze | Silica | $SiO_2$ |
| Radiolarian (single cell animal) | Radiolarian ooze | Silica | $SiO_2$ |

Although the minerals aragonite and calcite have the same chemical formula, aragonite is chemically less stable and dissolves more easily than does calcite. The depth at which a mineral begins to dissolve is called the "**compensation depth**" of the mineral (Figure 4-2). The compensation depth is not an absolute depth, but depends on several factors, such as the pressure and temperature of the water, as well as the supply of the soluble material. If enough soluble biogenic material is generated, the water will become saturated with dissolved material and dissolution will cease. Material will then begin to accumulate on the sea floor.

The aragonite compensation depth averages about 4000 m. Pteropod oozes are normally found only on the shallow peaks (<4000 m) of submarine ridges because pteropods (pelagic snails) are very sensitive to suspended terrigenous sediment and only live far from shore. Except for the ridge tops, most of the sea floor in the open ocean is below the aragonite compensation depth.

Calcite is stable to a depth of about 5300 m and coccolith, foraminifera, and ostracod oozes are normally found above this depth. Coccoliths are the remains of single-cell algae. Foraminifera are single-cell animals and ostracods are small shrimp-like organisms that secrete hinged shells. The oozes made by these three organisms are the most common oozes in the ocean basin.

Silica is stable to great depths and the distribution of silica oozes is mainly controlled by the distribution and productivity of silica-secreting organisms, not water depth. Diatoms are single-cell algae that thrive in subpolar environments. Radiolarians are single-cell animals and are most abundant in the tropics. Because these organisms normally reproduce more slowly than calcite or aragonite ooze-forming organisms, silica oozes occur mainly in areas below the aragonite and calcite compensation depth.

In the deep ocean regions not dominated by oozes, terrigenous, and authigenic (hydrogenic) sediments dominate. Terrigenous sediment consists mainly of **red clay,** actually brown in color and composed of wind-blown dust, cosmic material, and clay particles. Some anomalously large terrigenous grains may be present because of transportation by turbidity currents, ice, animals, and/or plants (Figure 4-2). **Ferromanganese nodules** are the most common authigenic grains present in the deep ocean. These may be so abundant as to blanket the sea floor. Phosphate nodules primarily occur on certain continental shelves.

## 4-5. BOTTOM SEDIMENT CHARTS

After bottom samples from an area have been collected and analyzed, the data are recorded on charts. Contiguous samples of like sediment types are mapped together as a unit to obtain a better picture of sediment distribution (Figure 4-3). Comparison with bathymetric charts aids in determining relationships among depth, sediment type, and a possible sediment source. Common abbreviations for sediment types used on navigation charts are given below. Where mixtures of two or more types of sediments occur, abbreviations may be separated by a slash (/).

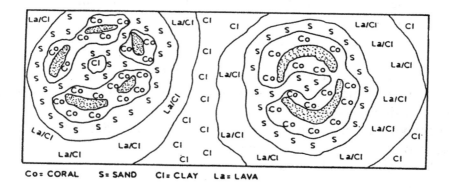

Co = CORAL    S = SAND    Cl = CLAY    La = LAVA

Figure 4-3.  Bottom sediment chart.

*Abbreviations used on navigational charts:*

| | | | | | | | |
|---|---|---|---|---|---|---|---|
| bk | black | Cl | clay | gl | Globigerina | | |
| Blds | boulders | Co | coral | Glac | glacial | | |
| br | brown | crs | coarse | gn | green | | |
| brk | broken | Di | diatom | Grd | ground | | |
| bu | blue | dk | dark | hrd | hard | | |
| Ca | calcareous | fn | fine | K | kelp | | |
| ch | chocolate | Fr | foraminifera | La | lava | | |
| Ck | chalk | G | gravel | lrg | large | | |
| lt | light | Oz | ooze | sft | soft | | |
| M | mud | P | pebbles | Sh | shells | | |
| Ml | marl | Pt | pteropods | sml | small | | |
| Mn | manganese | Qz | quartz | St | sticky | | |
| | nodules | Rd | radiolarian | Vi | violet | | |
| Ms | mussels | rd | red | Vol Ash | volcanic ash | | |
| Or | orange | Rk | rocky | yl | yellow | | |
| Oys | oysters | S | sand | | | | |

# EXERCISE 1.  BOTTOM SEDIMENT CHARTS

1. Sandy Harbor chart (Figure 4-4).
   a. Complete the bottom sediment chart by drawing the boundaries between the various sediment types.
   b. What is the dominant sediment shown on the chart? _____
   c. Is this a size class, origin or compositional term? _____
   d. Is there a relationship between sediment type and depth? (See bathymetry chart for Sandy Harbor, Figure 2-8.) If so, describe.


   e. What is the most probable type of sediment found at each of the following points on the chart?
      1. _____   2._____   3._____   4._____

2. Chart of a portion of the Pacific Ocean (Figure 4-5).
   a. Complete the bottom sediment chart by drawing the boundaries between the various sediment types.
   b. What type of sediment dominates the following areas (See bathymetry chart for a portion of the Pacific Ocean, Figure 2-9)?
      Shallow water around islands. _____
      Deep portion of ocean. _____
      Central depression surrounded by several islands near lower right-hand corner of chart. _____

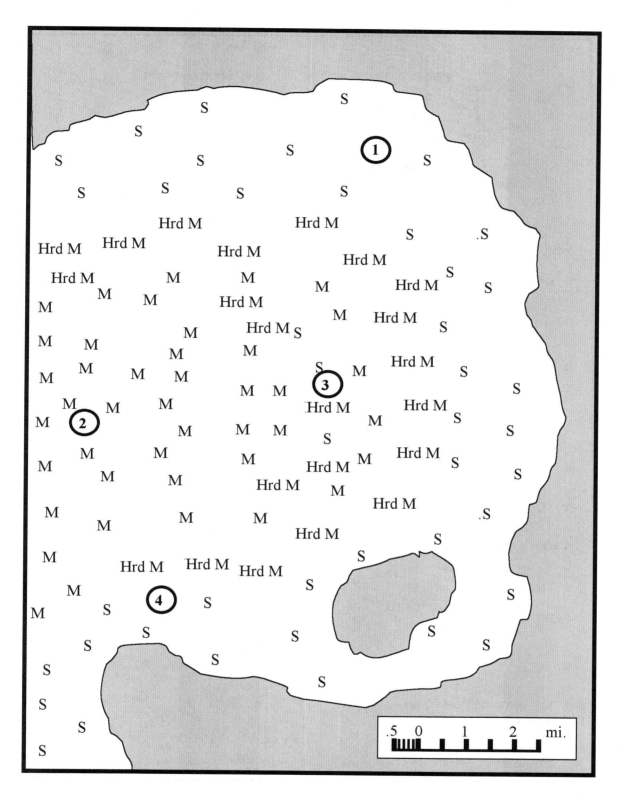

Figure 4-4.  Bottom sediment chart of Sandy Harbor.

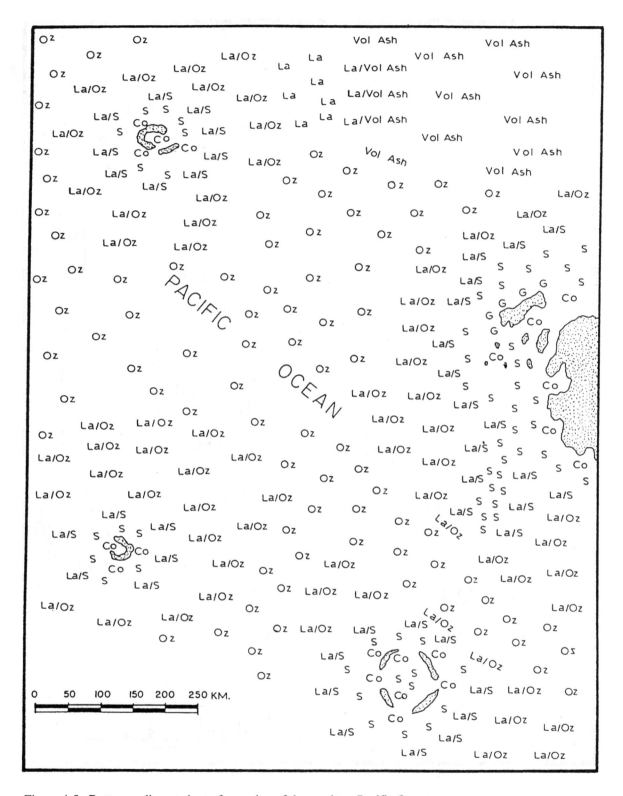

Figure 4-5.  Bottom sediment chart of a portion of the southern Pacific Ocean.

c.  Below are all of the abbreviations used on this chart. Write what each abbreviation means and state if the term indicates origin (terrigenous, biogenic or authigenic), size class or description.

Vol Ash _____ _____        *rd* _____ _____
Oz      _____ _____        La  _____ _____
G       _____ _____        S   _____ _____
Co      _____ _____        D   _____ _____

d.  Does sediment type appear to be related to depth? If so, describe.

e.  Considering the dominant bathymetric feature of the Pacific Ocean basin, what do the rock and sediment types on this chart and the general configuration of the sea floor around the island groups indicate as a probable origin for these islands?

f.  From your knowledge of compensation depths, would you expect the ooze on this chart to be mainly aragonite, calcite, silica or a mixture of all three? Why?

g.  Note the area of sand extending outward from the large island along the right edge of the chart. How can this be explained?

## EXERCISE 2.   OCEAN FLOOR FEATURES AND SEDIMENT TYPES

1.  On the following three profiles (Figure 4-6), identify the most probable type of sediment(s) expected in each of the lettered areas. On each profile, a continent is shown on the left and a broad ocean is to the right.

A. _____    G. _____    M. _____
B. _____    H. _____    N. _____
C. _____    I. _____    O. _____
D. _____    J. _____    P. _____
E. _____    K. _____
F. _____    L. _____

2.  The sediment accumulating in the deep ocean basin is primarily derived from shells of pelagic (floating) organisms. Explain why the sediment would be much thicker on the flanks of the Mid-Atlantic Ridge than on the ridge crest.

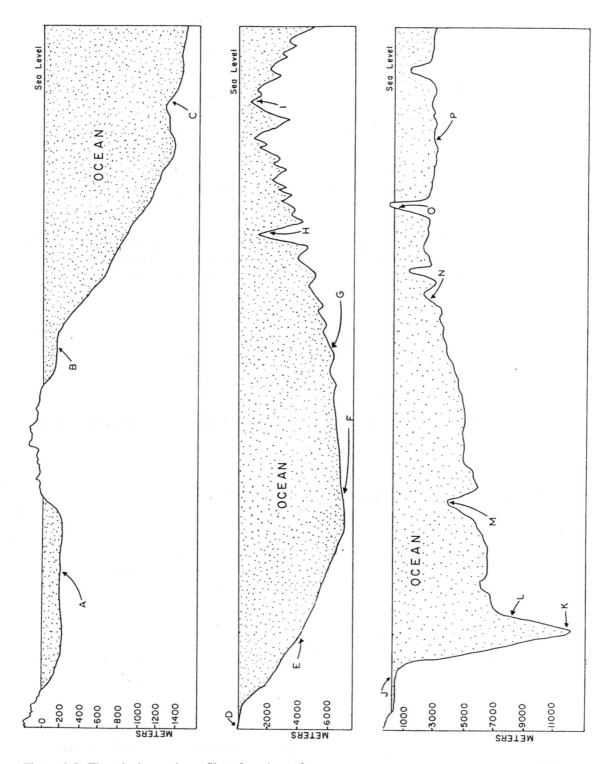

Figure 4-6.  Three bathymetric profiles of portions of an ocean.

3. How can the presence of granite (continental) cobbles and pebbles in the middle of the North Atlantic be explained?

## 4-6.  SEA FLOOR SEDIMENT THICKNESS

As soon as sea floor forms at the oceanic ridge crest, sediment begins to accumulate upon it. Rate of accumulation can vary depending on a number of factors such as:
1. Rate of organic productivity in the overlying waters.
2. Amount of wind-transported sediment from the land or fine particles from volcanic eruptions that settle from of the atmosphere.
3. Rate at which cosmic material falls to Earth.
4. Speed at which authigenic sediments form.
5. Quantity of water-transported sediments that sink to the sea floor.
6. Rate of sediment dissolution or erosion.
7. Amount of ice-rafted sediments released as icebergs melt.

In general, the farther a section of sea floor is from the oceanic ridge crest, the older the sea floor, the longer sediments have had to collect and the thicker the accumulation of sediment. Therefore, the three interrelated variables determining the amount of sediment found on the sea floor are the rate at which sediments collect, the age of the sea floor, and the distance the section of sea floor is from the oceanic crest. Given any two of these variables, it is possible to determine the third.

Example 1. If the average rate of ocean basin expansion is 3 cm/yr and a studied section of sea floor is 400 km from the crest, how thick should the sediments be if the average rate of accumulation is 2 cm/1000 yrs.?
  a. First determine the average sea floor spreading rate.
  Average sea floor spreading is the rate at which sea floor moves away from the rift valley on the ridge crest. It is equal to one-half the average rate of ocean basin expansion, the rate at which both sides of the oceanic ridge spread.

  Average ocean basin expansion rate = 3 cm/yr
  Average sea floor spreading rate = 3 cm/yr × 1/2 = 1.5 cm/yr

  b. Determine the age of the sea floor upon which the sediment has collected. This can be accomplished by calculating how long it has taken this section of sea floor to be moved 400 km from the ridge crest. Convert the 400 km into centimeters.

  400 ~~km~~ × 100 ~~m/km~~ × 100 cm/~~m~~ = $4 \times 10^7$ cm

To determine the time required for the sea floor to spread $4 \times 10^7$ cm at the rate of 1.5 cm/yr, divide the distance by the rate of spreading.

  $4 \times 10^7$ cm ÷ 1.5 cm/yr = $4 \times 10^7$ ~~cm~~ × 1 yr/1.5 ~~cm~~ = $2.6 \times 10^7$ yrs

The sea floor is 26,000,000 yrs old.

c. Sediment has been accumulating at the average rate of 2 cm/1000 yr. To calculate the expected thickness of sediment, multiply the time interval by the rate of accumulation.

$$2.6 \times 10^7 \; \text{yrs} \times 2 \; \text{cm}/10^3 \; \text{yrs} = 0.8666 \times 10^4 \; \text{cm} = 8666 \; \text{cm}$$

d. Convert the thickness into meters.

$$8666 \; \text{cm} \times 1 \; \text{m}/100 \; \text{cm} \cong 86.66 \; \text{m}$$

Example 2. Sediment has been collecting at the average rate of 3 cm/1000 yrs. A section of sea floor is covered by 60 m of sediment and is 200 km from the rift valley on the ridge crest. What was the average rate of sea floor spreading for this section of sea floor?

a. Determine the time interval required to collect 60 m of sediment if sediment was deposited at the average rate of 3 cm/1000 yrs. Convert the 60 m of sediment into centimeters of sediment.

$$60 \; \text{m} \times 100 \; \text{cm/m} = 6000 \; \text{cm}.$$

To determine the time required to deposit 6000 cm, divide the sediment thickness by the rate of accumulation.

$$6000 \; \text{cm} \div 3 \; \text{cm}/1000 \; \text{yrs} = 6000 \; \text{cm} \times 1000 \; \text{yr}/3 \; \text{cm} = 2 \times 10^6 \; \text{yrs}$$

If it has taken $2 \times 10^6$ yrs for the sediment to have accumulated, the sea floor must be at least $2 \times 10^6$ yrs old and it has required $2 \times 10^6$ yrs for the sea floor to have spread 200 km from the ridge crest.

b. To determine the average rate of sea floor spreading, convert the distance traveled into centimeters and then divide the distance by the time required to travel that distance.

$$200 \; \text{km} \times 1000 \; \text{m/km} \times 100 \; \text{cm/m} = 200 \times 10^5 \; \text{cm}$$
$$200 \times 10^5 \; \text{cm}/2 \times 10^6 \; \text{yrs} = 10 \; \text{cm/yr}$$

Example 3. A section of sea floor is covered by 600 m of sediment and is located 300 km from the rift valley in the ridge crest. Average sea floor spreading rate is 5 cm/yr. What is the average rate per 1000 yrs at which sediment accumulates on the sea floor?

a. To determine the age of the sea floor, calculate the time required for the sea floor to spread 300 km from the crest at 5 cm/yr. Convert the distance to centimeters and divide by the average spreading rate.

$$300 \; \text{km} \times 1000 \; \text{m/km} \times 100 \; \text{cm/m} = 3 \times 10^7 \; \text{cm}$$
$$3 \times 10^7 \; \text{cm} \div 5 \; \text{cm/yr} = 30 \times 10^6 \; \text{cm} \times 1 \; \text{yr}/5 \; \text{cm} = 60 \times 10^6 \; \text{yrs}$$

It has taken $60 \times 10^6$ yrs to collect 600 m of sediment.

b. To determine the average annual rate of sediment accumulation, convert the sediment thickness to centimeters and divide the amount of sediment by the time it took for the sediment to accumulate.

$$600 \text{ m} \times 100 \text{ cm/m} = 6 \times 10^4 \text{ cm}$$
$$6 \times 10^4 \text{ cm} \div 60 \times 10^6 \text{ yrs} = 0.001 \text{ cm/yr}$$

Convert to the average accumulation per 1000 years by multiplying the average annual rate of sediment accumulation by 1000/1000

$$1000/1000 \times 0.001 \text{ cm/yr} = 1 \text{ cm/1000 yr}$$

The average rate of sedimentation is obtained by dividing the total thickness of sediment deposited by the time over which the sediment has accumulated. If the rate of sedimentation has been fairly uniform, the average rate is a good approximation of the rate of sedimentation for any part of the sediment column. In contrast, if the rate of sedimentation has varied greatly, the average rate of sedimentation may not be truly representative of any section of the sediment column.

A **core** is a cylindrical sample of sediments or rock. Cores are usually a few inches in diameter and provide a relatively undisturbed sample of material. Various coring devices can sample deep-sea sediments down to the underlying volcanic layer. Using radiometric dating, it is possible to very accurately determine when the volcanic rocks formed and sediment accumulation began. Generally, it is assumed that sediment deposition started immediately after the sea floor rock formed. Fossils within a core allow subsequent ages of sediments to be determined. The average rate of sedimentation between each dated section of a core can then be determined by dividing the thickness of sediment in a section by the time interval during which that section was deposited.

## EXERCISE 3.  CALCULATIONS ON SEDIMENTS

1. If the average rate of sediment accumulation is 3 cm/1000 yrs, how long would be required for sediments to completely bury seamounts 1.5 km high and convert part of the oceanic ridge into a part of the abyssal plains? _____

2. The Atlantic Ocean began forming about 200 million years ago. If one applies the average sediment accumulation rate of 3 cm/1000 yrs, measured near the center of the ocean basin, to the entire basin, what is the maximum thickness of sediment that would be expected at the basin edge? _____
The actual thickness is much greater. How can this be explained?

3. Rates of sediment accumulation not only vary with time, but also with location across the ocean basin. In general, the rate of sediment accumulation is higher at the oceanic ridge and near the continent, but lower between them where the ocean is deeper. How can this be explained?

4. Given a constant average rate of sediment accumulation across the entire deep ocean basin, should the speed of sea-floor spreading influence the thickness of sediment that would accumulate on a section of sea floor? Why?

5. Below is a cross section of part of an ocean basin. Sedimentary cores will be collected at the points indicated (A-D). It is necessary to know the approximate maximum thickness of sediment in each of these locations to select the correct sampling device. Given the distance of each point from the rift valley, the average rate of sea floor spreading between each location and the average rate of sediment deposition, calculate the expected thickness of sediment at each location.

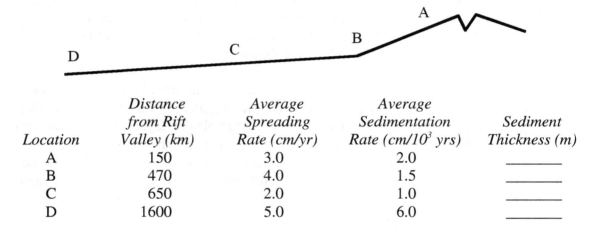

| Location | Distance from Rift Valley (km) | Average Spreading Rate (cm/yr) | Average Sedimentation Rate (cm/$10^3$ yrs) | Sediment Thickness (m) |
|---|---|---|---|---|
| A | 150 | 3.0 | 2.0 | _____ |
| B | 470 | 4.0 | 1.5 | _____ |
| C | 650 | 2.0 | 1.0 | _____ |
| D | 1600 | 5.0 | 6.0 | _____ |

6. Below is a cross section of part of an ocean basin. Sedimentary cores have been collected at the points indicated (A-D). Given the distance of each point from the rift valley, the average rate of sea floor spreading between each location and the thickness of sediment from each core, calculate the average rate of sedimentation at these localities.

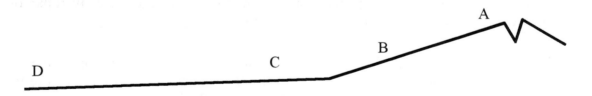

| Location | Distance from Rift Valley (km) | Average Spreading Rate (cm/yr) | Average Sedimentation Rate (cm/$10^3$ yrs) | Sediment Thickness (m) |
|---|---|---|---|---|
| A | 140 | 2.0 | _____ | 140 |
| B | 540 | 4.0 | _____ | 510 |
| C | 930 | 3.0 | _____ | 600 |
| D | 2910 | 9.0 | _____ | 4100 |

7. Below is a cross section of part of an ocean basin. Sedimentary cores have been collected at the points indicated (A-D). Given the distance of each point from the rift valley, the average rate of sedimentation and the thickness of sediment from each core, calculate the average rate of sea floor spreading between each location.

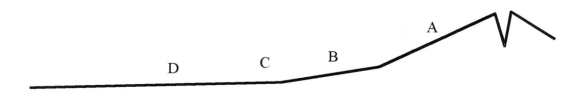

| Location | Distance from Rift Valley (km) | Average Spreading Rate (cm/yr) | Average Sedimentation Rate (cm/$10^3$ yrs) | Sediment Thickness (m) |
|---|---|---|---|---|
| A | 750 | _____ | 1.0 | 150 |
| B | 1200 | _____ | 3.0 | 1050 |
| C | 1440 | _____ | 3.0 | 1530 |
| D | 1640 | _____ | 15.0 | 13800 |

8. Below is a diagram (not drawn to scale) of a sedimentary core recovered from the deep sea floor. It is 4.07 km long. The top and bottom of the core are labeled. The core is oriented horizontally because that is the orientation in which they are normally studied. The volcanic rock just below the base of the core has been dated as being $50 \times 10^6$ yrs old. Fossils from within the core have been examined and they provided the additional dates, indicated in millions of years, on the core.

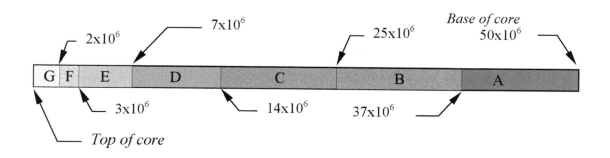

a. Determine the average rate of sedimentation for the entire core. _____
b. Determine the average rate of sedimentation for each segment of the core.

| Segment | Thickness (m) | Average rate of Sedimentation (cm/10³) |
|---------|---------------|----------------------------------------|
| A | 1950 | _____ |
| B | 1080 | _____ |
| C | 660 | _____ |
| D | 210 | _____ |
| E | 120 | _____ |
| F | 10 | _____ |
| G | 40 | _____ |

c. How representative is the average rate of sedimentation for the entire column compared to the average rate of sedimentation for each segment?

# Laboratory 5

# Salinity, Temperature and Turbidity

## 5-1. DENSITY STRATIFIED SYSTEM

The ocean is not a uniform body of salt water. In mid- to low latitudes (near the equator), the sea is vertically stratified into layers that differ in density because of varying salinity and temperature. These density differences prevent the layers from easily mixing. The greater the difference in density between two layers, the more stable and long-lived is the stratification (layering).

When two or more bodies of water of different density come together, each water mass flows to occupy a position determined by its density. The least dense flows to the top and the most dense sinks to the bottom. Intermediate densities occupy appropriate positions in between.

The density of water is determined by temperature and salinity. Density increases as salinity increases and temperature decreases. Generally, a warm water mass rests above a cold one and a less saline water mass sits upon a more saline one. Because both salinity and temperature vary in sea water, the density of a given body of water will depend on both of these factors. Thus, it is possible for a warm, more saline water mass to rest in equilibrium above a cold, less saline water mass if the warmer temperature makes the more saline water less dense. Similarly, a cold, less saline water mass can rest in equilibrium above a warmer more saline mass if salinity makes the warmer water mass denser than the colder one.

Turbidity currents are temporary water masses characterized by high density, resulting from a large amount of suspended particulate matter. Because turbidity currents are denser than surrounding water, they flow along the bottom, down the slope and toward the ocean floor. Turbidity currents are relatively short-lived phenomena because as soon as they cease flowing, the suspended sediment begins to settle and the density of the mass decreases quickly.

## 5-2. SALINITY IN THE SEA

**Salinity** is a measure of the quantity of salts dissolved in water and is expressed in parts per thousand, abbreviated **ppt**, or ‰. Parts per thousand means that in 1000 g of sea water, a certain number of grams are dissolved salts. For example, if the salinity is 20‰, this means that 20 g of a kilogram of sea water are salt and the remaining 980 g are pure water. Parts per thousand is similar to percentage except it is based on 1000 rather than 100. Thus, 20‰ is equal to $20/1000 \times 100\%$ or 2%. Thus, 2% of the salt water is salt and 98% is water.

Average salinity in the ocean varies from 33 to 37‰. In restricted areas, salinity may range from almost 0‰ (fresh water) to 72‰, the salinity at which **evaporites** (salts) begin to precipitate (crystallize) from solution. Local changes in salinity can result from evaporation, precipitation (rainfall, snow, etc.), inflow of fresh water from rivers or springs, formation or melting of sea ice, and diffusion. **Diffusion** is the slow, random motion of molecules and ions. Over long periods of time, diffusion will allow migration of dissolved salts from areas of higher salinity to those of lower salinity, thereby changing the salinity of both.

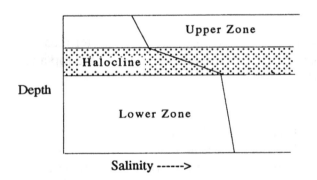

Figure 5-1. Change in salinity with depth.

Most changes in salinity occur at or near the surface because that is where the environment fluctuates more frequently, extremely and rapidly. Below the surface, salinity is generally stable. A graph plotting salinity versus depth demonstrates that salinity does not change uniformly with depth (Figure 5-1). Three distinct salinity zones can be recognized. The upper zone is marked by a gradual change (increase or decrease) in salinity with depth. Below the upper zone is the **halocline**, a zone of mixing that is characterized by a rapid change in salinity with depth. The lower zone is marked again by a gradual change in salinity. Thickness of the halocline varies with the amount of mixing between the upper and lower zones. The greater the mixing, the thicker is the halocline.

## EXERCISE 1. SALINITY AND WATER STRATIFICATION

Procedure

1. Install a partition in the middle of a clear tank or aquarium, as shown in Figure 5-2. Seal the sides of the partition with a thin layer of modeling clay.

2. Fill a container with room-temperature salt water and another with room-temperature fresh water. Add food coloring to each container. Use color combinations such as blue and yellow or red and yellow. Do not make the solutions too dark. Record the color of each solution (Figure 5-3).

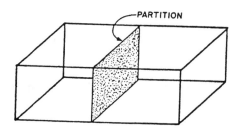

Figure 5-2. Container with partition.

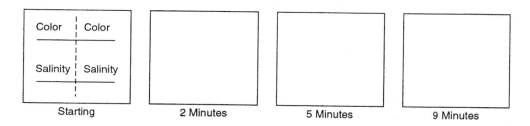

Figure 5-3. Changes in color as viewed through the tank's side.

3. Simultaneously pour one of the solutions on each side of the partition until both sides of the tank are filled to the same height.

4. Allow the solutions to calm. Quickly remove the partition, creating as little turbulence as possible. Through the side of the box observe what occurs to the solutions as the partition is removed. It may help to hold a sheet of white paper behind the tank.

5. In the area provided, draw the changes observed at the times suggested (See instruction No. 6). Label colors and zones present (Figure 5-3). Measure and record the thickness of each layer.

6. After 2.5 min, blow moderately hard across the water at a low angle to the surface along the length of the tank for about 15 sec. Observe what happens through the side of the tank.

7. Blow across the surface again after 9 min.

8. After 12 min, plot the relative change in salinity (colors) with depth. Label zones present (Figure 5-4). Graph relative change in salinity (color) versus depth at 12 min.

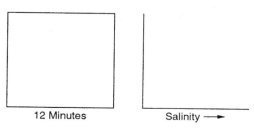

9. Dispose of solutions when finished.

Figure 5-4. Changes in salinity after 12 min.

Questions

1. Briefly describe what occurred when the partition was removed.

2. What produced the motion described above?

3. Where would the mixing of fresh water and salt water occur in nature?

4. How is the zone of mixing, the halocline, identified in this experiment?

5. Describe what occurred when you blew across the surface.

6. Does the zone of mixing become thicker or thinner after blowing across the surface? Why would you expect this?

7. From 2.5–9 min, the zone of mixing gradually should be thickening, although you may not be able to easily observe this. What process is causing this slow mixing?

8. Does the zone of mixing appear thicker or thinner after blowing across the surface the second time (after 9 min)?

9. How permanent does stratification caused by differences in salinity appear to be?

10. Is the halocline a zone of rapid increase or decrease in salinity?

**General Questions**

1. The surface layer and the lower layer in two different tanks both differ in salinity by 20‰. In one container the halocline is 3 cm thick and in the other it is 15 cm thick.
   a. How do the two haloclines differ in the rate of change of salinity?
   b. Which one would you expect to be more easily mixed by blowing across the surface of the water? Why?

## 5-3.  TEMPERATURE AND WATER STRATIFICATION

The Sun is the major source of heat for the oceans, but it directly heats only the surface layer. The deep ocean maintains a constant temperature below zero is some areas. Because the lower latitudes receive the most direct solar radiation, tropical surface water is considerably warmer than the rest of the ocean. Near the poles, where solar radiation is most diffuse, the surface water is only slightly above freezing. Tropical oceans display very strong thermal stratification all year because the warm surface water is much less dense than the deeper cold water. No stratification exists in the polar regions because the deep and surface waters are nearly **isothermal**, the same temperature. In the mid-latitudes, stratification is seasonal.

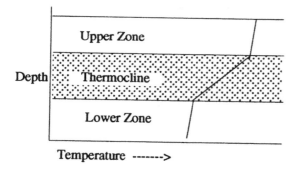

Figure 5-5.  Changes in temperature with depth.

A graph plotting temperature versus depth shows that temperature does not decrease uniformly from the surface to the ocean floor in the tropics and subtropics (Figure 5-5). Three distinct temperature zones can be recognized. The warm **surface zone** is characterized by a gradual decline in temperature with depth. Beneath the surface zone is the **thermocline**, a zone of mixing characterized by a rapid change in temperature with depth. The **bottom zone** is marked by very cold water and a gradual temperature decline with depth.

## EXERCISE 2.  TEMPERATURE AND WATER STRATIFICATION

Procedure

1-4. Follow the instruction as in Exercise 1, except use warm and cold freshwater. Graph the change in temperature (color) with depth at the times requested (Figure 5-6).

5-7. Same as in Exercise 1.

8.   Allow the solutions to become isothermal. Blow across the surface and observe the changes.

9.   Dispose of solutions.

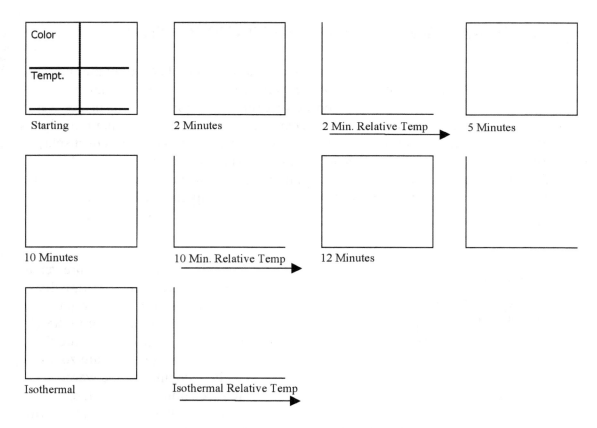

Figure 5-6.  Changes in color with time as viewed through the side of the tank.

Questions

1.  Briefly describe what occurred when the partition was removed.

2.  What produced the motion described above?

3.  Where could the mixing of warm and cold salt water occur in nature? Warm and cold fresh water?

4.  How is the thermocline identified in this experiment?

5. Describe what occurred when you blew across the surface the first time.

6.  Do the solutions appear to mix more easily as they become isothermal? When they are isothermal? Why?

7. How permanent does stratification from temperature appear to be in this experiment?

8. Why is thermal stratification more permanent in the tropics and less permanent in the temperate (midlatitude) regions?

## 5-4. DENSITY AND WATER STRATIFICATION

Density of sea water is mainly controlled by salinity and temperature. As seen from the previous two experiments, density increases with increasing salinity and decreasing temperature. Because of variations in temperature and salinity in the ocean, density stratification extends from the equator into the temperate regions (Figure 5-7).

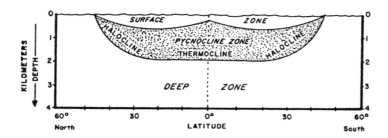

Figure 5-7.  Basic ocean stratification.

In the tropics, surface waters are less dense because the high amount of rainfall lowers salinity and the year-round tropical climate warms the water. In the subtropics, around 30°N/S, the very warm surface waters are less dense than the waters below, despite increased evaporation and higher salinity. In the temperate region, surface waters are less dense despite being relatively cool, because of increased precipitation (rainfall, snow, etc.). In the polar regions, surface and deep waters have nearly the same density.

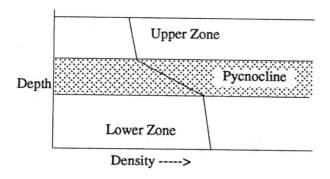

Figure 5-8.  Change in density with depth.

A graph plotting density versus depth shows three layers in much of the ocean (Figure 5-8). In the upper zone, density increases gradually downward. The underlying **pycnocline** is characterized by a rapid increase in density with depth and the lower zone displays a gradual increase in density with depth.

In the ocean, the upper density layer is called the **surface zone** (Figure 5-7). Water in this zone is characterized as being warmer and less dense with variable salinity because of precipitation and evaporation. The surface zone extends downward to a maximum depth of about 100 meter at about 30° north or south of the equator. It contains only 2% of the ocean volume and is well mixed by winds, waves and currents. Much of the organic production in the ocean occurs within the surface zone.

The pycnocline is characterized by rapidly increasing water density with depth because of changes in temperature or salinity. It is thickest (>1.7 km) in the tropics and thins poleward. Its base curves upwards and extends to the surface at about 45° north and south latitude, depending on the season. The pycnocline coincides with the halocline in the temperate region, the thermocline in the subtropics and a combination of halocline and thermocline in the tropics.

The **deep zone** is extremely cold water and increases in density with depth because of increasing pressure and declining temperature. It contains over 80% of all ocean water, is homogeneous and extremely stable. The water of the deep zone extends from the base of the pycnocline to the ocean bottom. In polar and subpolar areas, where no pycnocline exists, waters of the deep zone extend to the surface.

## EXERCISE 3.  A MULTIPLE-LAYER SYSTEM

The great thickness of the pycnocline in the tropics effectively isolates the deep layer from surface phenomena. The isolation of the deep layer can be studied in a multiple-layer system.

Procedure

1.  Obtain an additional partition. Use the two partitions to divide the tank into thirds and seal the sides of the partitions with clay.
2.  Fill one container with hot fresh water, a second with room-temperature salt water and a third with cold salt water. Add a different food coloring to each as before.
3.  Simultaneously pour each solution into a section of the tank and try to fill the tank almost to the top of the partitions. For the system to work best, pour the room-temperature salt solution into the middle third.
4.  Allow the solutions to calm and then remove both partitions at the same time, attempting to create as little turbulence as possible. Through the side of the tank, observe what occurs. In the area provided, draw the changes seen at the times suggested (Figure 5-9). Label colors and zones present.
5.  After 2.5 min, blow moderately hard across the water surface along the length of the container at a low angle to the surface for about 20 sec. Observe what occurs to the lowermost solution.
6.  After 10 min, try blowing harder and longer and observe any changes.
7.  Dispose of the solutions.

Questions

1. What occurred to the bottom solution as you blew across the surface the first time?

2. What occurred to the bottom solution as you blew across the surface the second time?

3. Why might a difference be expected between the first and second time you blew across the surface?

4. How protected was the lowermost layer in the multiple-layer system?

5. In the ocean, why would one expect the thermocline to be more important in forming the pycnocline in some areas, but the halocline to be more important in other regions?

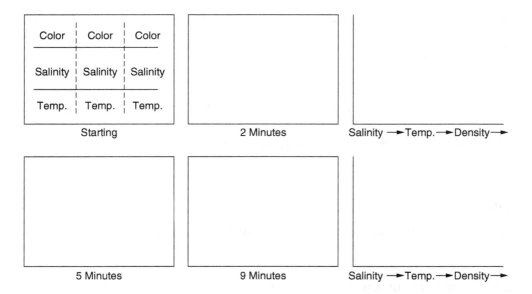

Figure 5-9. Changes in color with time.

## 5-5.  TURBIDITY CURRENTS

Turbidity currents are temporary, flowing water masses characterized by extremely high density resulting from a large amount of suspended sediment. Because turbidity currents depend on sediment suspension, they exist only as long as there is sufficient turbulence. This is directly related to the steepness of the slope down which the current flows: the steeper the incline, the faster the flow and the greater the turbulence. As the slope declines, turbulence decreases and the sediment begins to be deposited in accordance with the Hjulstrom Diagram: largest particles first and smallest last. In the ocean, turbidity currents commonly begin on the shelf, flow rapidly down the slope, and die on the continental rise and the abyssal plains.

## EXERCISE 4.  TURBIDITY CURRENTS

Procedure

1-5.  Same procedure as for Exercise 1 of this lab, but use cold clear fresh water and a turbid solution. Mix the turbid solution well. Do not add color to either solution. Draw and graph change in turbidity (Figure 5-10).

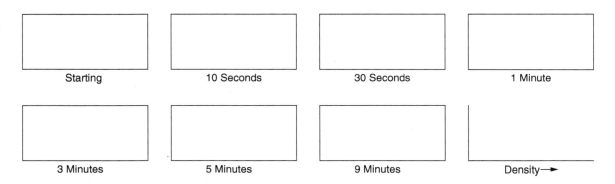

| Starting | 10 Seconds | 30 Seconds | 1 Minute |

| 3 Minutes | 5 Minutes | 9 Minutes | Density ⟶ |

Figure 5-10.  Changes in turbidity and density as viewed through the side of the tank.

Questions

1. Briefly describe what occurred when the partition was removed.

2. What produced the motion described above?

3. What occurred to the turbid zone when you blew across the surface?

4. What happens to the dense turbid zone with time?

5. How permanent does stratification from turbidity appear to be? (Once the sediment is deposited, it is no longer a water layer and should not be considered in terms of stratification.)

## EXERCISE 5. SUMMARY QUESTIONS

1. Rank temperature, salinity and turbidity in order of producing the most stable and long-lived stratification in these exercises. Why are some more stable than others?

2. Which variable would be most likely to produce the most stable and long-lived stratification in nature? Why?

3. How do the changes in turbidity with time differ from the changes observed for salinity and temperature?

## 5-6. METRIC (CELSIUS)-ENGLISH (FAHRENHEIT) TEMPERATURE CONVERSIONS

The **Fahrenheit** temperature scale was developed by Gabriel Daniel Fahrenheit in Holland in 1724. (The so-called "English" temperature scale has nothing to do with England or an Englishman.) The exact origin of the scale is uncertain and several different stories exist. One is that he placed zero degrees Fahrenheit at the lowest temperature he measured during the winter of 1708–1709 and 100° as the temperature of his body. He then divided the scale into 12 divisions and each of these into eight units. On the Fahrenheit scale, fresh water freezes at 32°F and boils at 212°F.

The **Centigrade** or **Celsius** temperature scale was developed by Anders Celsius in Sweden in 1742. He used the freezing point of fresh water as 0°C and the boiling point of fresh water as 100°C. Because the temperature difference between the two reference points is 100 units, the scale was initially called **centigrade** (centi = 100; grade = unit). Because it is based on a multiple of 10, this scale is considered part of the metric system.

Conversion between single degrees Fahrenheit and Celsius is very simple: 1°C = 1.8°F and 1°F = 0.56°C. When converting temperatures between the two scales, however, it must be remembered that 0°C is equal to 32°F. This means that when converting to

Fahrenheit from Celsius, it is necessary to add 32°. Similarly, when converting Fahrenheit to Celsius, you must first subtract 32°.

$$°F = (°C \times 1.8) + 32° \quad \text{or} \quad °F = 9/5°C + 32°$$
$$°C = (°F - 32°)0.56 \quad \text{or} \quad °C = 5/9(°F - 32°)$$

## EXERCISE 6.  TEMPERATURE CONVERSIONS

1.  Convert each of the following into the temperature of the other system.
    a.   1°C = _____°F        c.  −10°C = _____°F        e.   31°F = _____°C
    b.  27°C = _____°F        d.     1°F = _____°C        f. −25°F = _____°C

2.  Convert each of the following into the temperature of the other system.
    a.  A change of 05°C = ____°F
    b.  A change of 30°F = ____°C

3.  At what temperature would thermometers using the Fahrenheit or Celsius scale both read the same number of degrees? _____

# Laboratory 6

# Ocean Circulation

## 6-1. OCEAN CURRENTS

An **ocean current** is a discrete flowing mass of seawater. Its motion is induced by wind, gravity or density differences with adjacent water masses. As currents flow, their paths are deflected by the Coriolis Effect. Two major current systems exist within the ocean: surface currents and thermohaline currents. **Surface currents** are mainly wind generated, restricted to the upper layer of the ocean and characterized by the horizontal movement of water. They primarily flow parallel to latitude, except where deflected to the north or south by a landmass.

**Thermohaline currents** are generated by density differences between water masses. They usually originate as water becomes denser because of increasing salinity or decreasing temperature. The dense water mass sinks and then flows horizontally outward in all unobstructed directions. The major oceanic thermohaline circulation system transports cold, dense water from the polar and subpolar regions toward the equator along the sea floor. Other forms of thermohaline circulation occur in estuaries, where fresh water flows across sea water, and in semi-isolated seas, where highly saline water sinks as it enters the ocean and then flows outward across the ocean basin.

## 6-2. CORIOLIS EFFECT

Coriolis Effect refers to the relative deflection of an object from a straight path as it moves across a rotating surface. As a surface rotates about an axis, all points have the same angular rotational speed (angle/time), but linear speed (distance/time) increases with distance from the axis. For example, as a disk rotates 45° in a minute, regardless of the distance from the axis, all points on the disk rotate at a speed of 45°/min. In contrast, the linear distance traveled by a point on the disk varies with its distance from the axis. Adjacent to the axis, a point moves a nanometer per minute, but 1 m away from the axis, a point's linear speed would be about 78 cm/min. It is the difference in linear rotational speed that produces the Coriolis Effect.

As viewed from the North Rotational Pole, Earth rotates counterclockwise (Figure 6-1). Every object on the Earth possesses a linear rotational speed determined by its distance

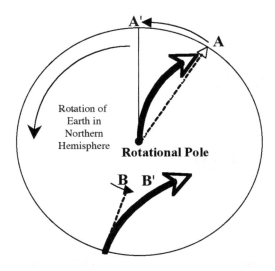

Figure 6-1. Coriolis Deflection. A—original location of destination. A′—location of A when plane reaches equator. B—original location of destination. B′—location of B when plane reaches latitude of B. Dashed line is path flown by the plane. Bold line is apparent path of flight. The plane appears to veer to the right of direction of travel in the Northern Hemisphere.

from the rotational pole. If an object, such as a plane, takes off from the pole and flies southward towards some point on the equator, it retains its original linear rotational speed even as it flies over parts of Earth's surface with greater linear rotational speed. If the plane flies in a straight course towards its destination, the ground below the plane, including its destination, will be moving eastward with a greater linear speed. By the time the plane reaches the equator, its destination will be some distance to the east. To an observer it will appear as though the plane is following a path that curves to the right of a line drawn between the launch site and destination. Again, the plane has flown in a straight line, but the Earth has rotated beneath it. The deflection is apparent and not real. No force pushed the plane to the right.

Similarly, as a plane travels from the equator northward towards some point, the plane maintains its linear rotational speed as it flies over parts of Earth's surface with a slower linear rotational speed. As a result, the plane moves eastward faster than its destination. To an observer it would appear as though the plane is following a path that curves to the right of a line drawn between the launch site and destination.

Because the Earth appears to rotate counterclockwise in the Northern Hemisphere, Coriolis causes moving objects to be deflected to the right of their direction of travel. As viewed from the South Rotational Pole, Earth appears to rotate clockwise. Coriolis deflects moving objects to the left of their direction of travel in the Southern Hemisphere.

## EXERCISE 1.  CORIOLIS EFFECT

Procedure A

Push the point of a thumbtack through a length of adhesive tape so that the head of the tack is on the adhesive side. Stick the tape to the desk with the tack pointing upwards. Place the center of a round sheet of paper (or paper plate) over the tack and push the point of the thumbtack through. The point of the thumbtack represents Earth's rotational axis and the paper represents Earth's surface.

1. Mark a point on the outer edge of the paper. This represents the destination of the plane. Place a coin on the desk next to this mark so that the paper can be rotated without moving the coin. The coin identifies the original location of the destination.
2. Draw a straight line across the paper from the axis to the destination.
3. Place the point of your pencil adjacent to the axis on the line you have just drawn.
4. As your lab partner begins to rotate the paper counterclockwise about the axis, draw a line from the axis to the coin. Try to ignore the paper's movement and draw straight across.
   a. Describe the relative relationship between the lines drawn in steps B and D.
   b. Why are the two lines not the same?
5. Using a new sheet of paper, repeat steps A through C.
6. As your lab partner begins to rotate the paper clockwise about the axis, draw a straight line from the axis to the coin.
   a. Describe the relative relationship between the lines drawn in steps B and F.
   b. Why are the two lines not the same?
   c. How do the lines drawn in steps D and F differ and why?
7. Mount a new sheet of paper and repeat steps A through C.
8. For each of the following steps, rotate the plate at the same speed. As your lab partner begins to rotate the paper counterclockwise about the axis, rapidly draw a straight line from the axis to the coin. Label this line "fast." Repeat the process, but draw the next line at about half the speed. Label this line "medium." Repeat the process but draw the final line slowly. Label this line "slow."
   a. Does a fast moving object or slow moving object display greater Coriolis displacement? Why?
   b. Would you expect Coriolis to be stronger in the polar or equatorial regions? Why?

Questions

1. If all of the Earth rotates in the same direction, why does it appear to rotate counterclockwise when viewed from the North Pole and clockwise when viewed from the South Pole?

2. How would Coriolis be expected to alter the path of an object if it crossed the equator?

3. Would an object traveling due east or west be expected to show Coriolis deflection? Why?

4. What factors would be expected to determine the amount of Coriolis deflection on a planet and why?

5. If the Earth revolved in the opposite direction, how would Coriolis be different?

## 6-3.  WIND AND SURFACE CIRCULATION

**Wind** can be defined as the movement of air parallel to Earth's surface. **Surface winds** are those winds that are in contact with Earth's surface. In contrast, **high altitude winds** occur high in the atmosphere and have little direct impact on surface phenomena. As surface winds blow across the ocean, friction drags the surface water in the general direction the air is moving. Adjacent water flows in to replace the water removed and also is transported by the wind. In areas of prevailing winds, a constant flow of water is established. Large-scale, unidirectional flow of water produced by the wind is called a **wind-generated current** (Figure 6-2). Speed of a wind-driven current is greatest at the surface and decreases with depth because energy transfer downward in the water column is inefficient. Where currents collide with land, they are deflected and flow along the landmass, as a **boundary current**.

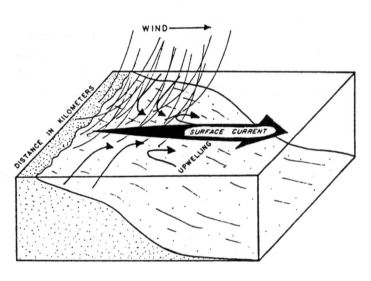

Figure 6-2. Formation of wind-generated current and upwelling.

Each hemisphere has three major wind belts and three windless zones between them (Figure 6-3). When the Sun is directly above the equator, the major surface winds are **Trade Winds** (0° to 30°N/S), **Westerlies** (30° to 60°N/S) and **Polar Easterlies** (60° to 90°N/S). Winds are named for the direction from which they originate. For example, westerlies blow from the west to the east. "Windless" zones are created where either cold, dry air sinks to Earth's surface forming a zone of **high pressure** or warm, moist air is raised from Earth's surface forming a zone of **low pressure**. There are no winds in areas of high pressure or low pressure because the air is moving vertically, not parallel to the surface. The windless zones are the **Doldrums** (low pressure, 0°), **Horse Latitudes** (high pressure, 30°N/S), **Subpolar Lows** (60°N/S) and **Polar Highs** (90°N/S). If there were no Coriolis deflection, the major winds would blow directly north and south, parallel to the lines of longitude. Coriolis deflects the winds by about 45° to the right of direction of travel in the Northern Hemisphere and to the left of direction of travel in the Southern Hemisphere. In the Northern Hemisphere, Trade Winds blow toward the southwest and the Westerlies blow toward the northeast. In the Southern Hemisphere the Trade Winds blow toward the northwest and the Westerlies blow toward the southeast.

Water propelled by the wind is also affected by Coriolis (Figure 6-3). Currents are deflected 45° to the direction of the winds. Thus, wind-driven ocean currents flow east and west, per-pendicular to the lines of longitude and parallel to the lines of latitude. In the North Atlantic, the Trade Winds produce the **North Equatorial Current** that flows westward from Africa to South America. Similarly, the Westerlies produce the **North Atlantic Current** that flows eastward from North America toward Europe. Ocean currents—in contrast to winds—are identified by the direction towards which they travel. The North Equatorial Current is a westerly current and the North Atlantic Current is an easterly current.

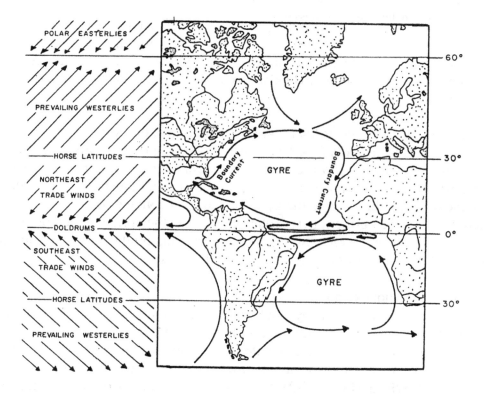

Figure 6-3. Major winds, windless zones and the Atlantic Ocean gyres and countercurrents.

When wind-generated currents impinge against a land mass, they become **boundary currents** and flow parallel to the land until they are diverted by other winds (Figure 6-3). In boundary currents, water is flowing down a slope (Figure 6-4). For example, the Trade Winds pull water away from the west coast of Africa, forming a slight depression in the ocean surface. This water accumulates along the coast of South America forming a low mound. To the north, the Westerlies pull water away from the east coast of North America, forming a depression and pile the water against the west coast of Europe. Gravity causes the water to flow down the sea surface from the mounds towards the depressions, thereby forming the boundary currents. In the North Atlantic, the downslope flow of water on the west side of the basin is called the **Gulf Stream** and on the east side of the basin it is called the **Canary Current**. In each ocean in each hemisphere, the wind-driven and boundary currents combine to form a circular path called a **gyre** (Figure 6-3). Because Coriolis deflection is reversed in northern and southern hemispheres, the resulting gyres rotate in opposite directions: counterclockwise in the Southern Hemisphere and clockwise in the Northern Hemisphere.

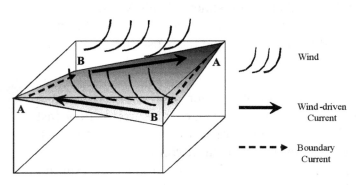

Figure 6-4. Boundary currents. Westerlies and trade winds drive water against the continents forming mounds (A) on one side of the ocean basin and depressions (B) on the other. Boundary currents flow down slope from the mounds to the depressions.

In the doldrums between the northern and southern gyres, the **North** and **South Equatorial Countercurrents** flow from west to east (Figure 6-3). Because there are no strong winds in this area of low pressure, water from the mound is free to flow eastward into the depression on the opposite side of the ocean basin. Equatorial countercurrents rotate in the opposite direction to the gyre and adjacent countercurrent. In the Atlantic and Pacific oceans the northern gyres rotate clockwise, the northern equatorial countercurrents turn counterclockwise, the southern equatorial countercurrents spin clockwise and the southern gyres rotate counterclockwise.

Surface currents do not extend to great depths, but can influence deeper layers of water in several ways. In areas where the pycnocline is relatively shallow, surface currents can induce a weak current in the denser waters of the pycnocline. Where strong persistent winds develop, currents move surface water away from the land and denser underlying water from the pycnocline may be drawn upward to replace the departing surface water. This is called **upwelling** (Figure 6-2). Under special conditions of prolonged upwelling of nutrient rich water, **phosphorite nodules** can form. In contrast, if winds drive water toward a landmass, surface waters will accumulate, sink and depress the pycnocline to a greater depth. This is called **downwelling**.

# EXERCISE 2.  WIND-DRIVEN WATER CURRENTS

Procedure A

1.  Obtain a rectangular tank that has a transparent bottom.
2.  On a sheet of white paper at least as large as the bottom of the tank, measure 2 cm distances along the length of the paper and draw parallel lines across the width of the paper. Number the lines consecutively: 2, 4, 6, etc.
3.  Place the container over the paper so that the numbered lines are clearly visible.
4.  Follow instructions as in Laboratory 5 to establish a layered fresh water-salt water system. Use as little clay as possible to seal the partition. On one side, place light blue salt water and on the other side, place the warm, colorless fresh water. Remove the partition and allow the solutions to calm.
5.  Blow vigorously along the length of the tank at a low angle to the water surface for about 15 sec. Have someone observe through the side of the tank that is parallel to the direction in which you are blowing.

Questions

1.  Briefly describe the changes observed as you blew across the water.


2.  What produced the changes observed?


3.  If the two ends of the tank are considered to be edges of continents, the water between them an ocean, and your breath the wind, what phenomena were observed as you blew across the water?


4.  Was the circulation established in the surface layer or thermohaline?  Why?


5.  What type of authigenic sediment might form near the side where you were blowing if this were a real ocean and this area a continental shelf?

Procedure B

1.  Measure and record in the spaces below the width of the container and the thickness of the surface layer (water surface to pycnocline) and bottom layer (base of pycnocline to bottom).

Width of container:        _____cm
Surface layer thickness:   _____cm
Bottom layer thickness:    _____cm

2. Blow gently across the water along one of the long sides of the container for about 40 sec. Wait about 15 sec, or until all waves have subsided, and then add 1 drop of red coloring to one end of the container and a drop of green coloring to the other end. Try not to vibrate the system and do not blow across the surface again. With a watch, time how long it takes for the coloring from either drop to travel 10 cm. Use the lines on the sheet of paper below the container to measure the 10 cm distance. Record the time and determine the speed of the water in centimeters per second.

   Coloring took _____ sec to travel 10 cm at _____cm/sec.

3. In the boxes on the following page (Figure 6-5), draw the distribution of the colors as viewed from above at the times suggested. Periodically, view the color distribution from the side. Because of the small size of the tank, Coriolis deflection is minimal.

Questions

1. What was the source of energy that set the water in motion?

2. How far down did the drops of food coloring sink? Why?

3. The tank is not actually analogous to the ocean because of its small size, but it does approximate circulation in a deep, elongated sea. Initially there should have been one large gyre. What happened when the "wind" ceased to blow?

4. Why do adjacent gyres rotate in opposite directions in the container?

5. Do the colors spiral inward or retain their position relative to the outside of the gyres?

6. If each particle of coloring is considered a drop of water and the coloring slowly spirals inward, then the water itself is moving toward the center of the gyre. Where does the water from the center of the spiral go?

7. How does the distribution of colors in the side view change with time?

8. How evenly distributed are the two colors at the end of 15 min?

9. Does the actual surface of the water or the surface layer continue to circulate longer? Why?

15 Seconds  45 Seconds  90 Seconds  2 Minutes 15 Sec.  3 Minutes

4 Minutes  5 Minutes  8 Minutes  10 Minutes  12 Minutes

14 Minutes  16 Minutes  18 Minutes  20 Minutes  22 Minutes

Figure 6-5. Boxes in which to plot flow of color in Exercise 2, Procedure B.

Procedure C

1.  Add salt to a small volume of red coloring until the salt ceases to dissolve, or obtain a highly saline dark red solution from your teacher. Using a straw or pipette, inject a few drops of the salty red coloring *into* the top of the dense saline layer. Blow gently across the water surface along one side of the container for about 40 seconds.

2.  Immediately add a drop of blue coloring to the surface layer directly above the drop of red. With a watch, time how long it takes for the blue coloring to travel 10 cm and the red coloring 4 cm. Record these times below. Determine their speed per second. Observe the solutions as they circulate for about 5 min.

> Blue took _____ sec to travel 10 cm at _____ cm/sec
> Red took _____ sec to travel 4 cm at _____ cm/sec

Questions

1.  Is there a delay between the first "wind" striking the surface and movement of the red coloring on the pycnocline? Why?

2.  After 5 min, has the red coloring shown evidence of water movement in the deeper layer? After 10 min?

3.  From this experiment, would the surface currents in the ocean be expected to exert much of an influence on bottom water circulation? Why?

Procedure D

Take a round bowl or glass and partially fill it with water. With a spoon, stir the water rapidly in a circle. Remove the spoon and immediately observe the water's surface.

Questions

1.  Is the surface flat, concave or convex?

2.  In the ocean, Coriolis creates a mound of water inside the gyre. In the container we have created a circular current similar to a gyre. Yet, the center is depressed and not formed into a mound. Why? What forces are involved?

## 6-4.  THERMOHALINE CIRCULATION

**Thermohaline circulation** is generated by density differences between water masses. It can produce both horizontal and vertical flow. Water density increases as water becomes cooler or more saline. If density increases sufficiently, the water sinks to a level such that all water below it is denser and above it is less dense. The water mass then flows outward in all unobstructed directions, sinking below less dense and rising above more dense water masses it encounters.

Several forms of thermohaline circulation occur in the ocean. In subpolar to polar regions, cold, dense water descends and slowly flows towards the equator. These water masses may be more saline than regular sea water because salts are concentrated in them as sea ice forms. As these currents slowly flow, they mix with adjacent water masses and eventually lose their identity.

Excess evaporation in a restricted sea can produce dense, saline water that sinks to the appropriate density level as it flows out into the ocean. Circulation through the Strait of Gibraltar, the entrance to the Mediterranean Sea, is an example of this form of thermohaline circulation.

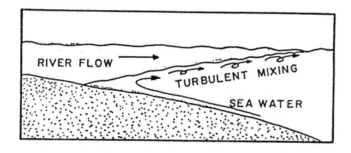

Figure 6-6. Estuarian circulation.

Thermohaline circulation also occurs in estuaries (Figure 6-6). As a river flows into an estuary, or narrow bay, the less dense fresh water forms a wedge that displaces the denser sea water. The wedge thins seaward as the fresh water spreads outward and slowly mixes with the sea water. Friction between the outflowing fresh water and the underlying salt water establishes **estuarian circulation**. This can generate upwelling within the estuary.

# EXERCISE 3. THERMOHALINE CIRCULATION (ESTUARIAN CIRCULATION)

Procedure

1. Place an aquarium so that one end is a few centimeters higher than the other and the lower end can drain into a sink (Figure 6-7). Fill the aquarium with clear salt water. Add a drop of salted red coloring to several points along the length of the aquarium.
2. With a hose placed at the raised end of the aquarium, slowly allow fresh water to flow across

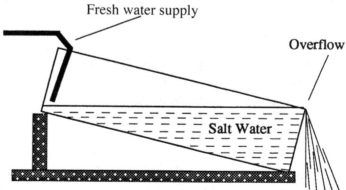

Figure 6-7. Estuarian circulation apparatus.

the salt water layer and overflow the lower end of the aquarium. Note what happens to the red coloring.

3. Add three drops of yellow coloring to the fresh water near the upstream end of the aquarium but away from where the water is flowing directly in. Observe what occurs.

Questions

1. Does much mixing of the water bodies appear to occur? Why?

2. Is the surface between the two water masses smooth or irregular? Describe.

3. Which flows faster, the surface or the deeper water?

## 6-5. SUBMARINE WATERFALLS

A thermohaline current flows outward across the ocean basin in all directions unless blocked by a landmass, submarine ridge or denser body of water. Where a submarine ridge blocks the flow of dense sea water, the dense water accumulates behind the ridge and may eventually overflow, cascading down the barrier as a **submarine waterfall**. At least four submarine waterfalls exist in the Atlantic Ocean. They are at Rio Grande Rise, Ceara Abyssal Plain, Denmark Strait Cataract and Gibraltar Strait.

Because the density of the falling water may be only slightly greater than the water through which it is falling, the water falls slowly and with less force than water falling through air. Yet, there is sufficient turbulence to promote mixing between water masses.

## EXERCISE 4.  SUBMARINE WATERFALLS

Procedure A

1. Partially fill an aquarium with clear, room temperature fresh water.
2. Completely fill a smaller and shorter container with room temperature fresh water. Add several few drops of yellow coloring and stir. Holding a piece of cardboard over the top of the container, carefully place it upright into the aquarium so that its top is well below the surface of the water. Incline the container by placing a wedge under the bottom, so one edge of the mouth is lower and overhangs the side (Figure 6-8). Slowly slide the cardboard from the top of the container. The aquarium represents the ocean basin and the container is an area surrounded by barriers. The yellow water has almost the same density as the surrounding water and represents water that initially filled the confined basin.

3. Fill a large container with very cold fresh water and add blue coloring. Place the container so that it is higher than the aquarium.

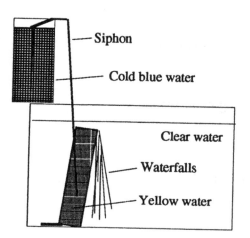

4. Fill a length of small diameter plastic tubing with clear room temperature water. Place a finger firmly over each end of the tube to form an air-tight seal. Put one end of the tube into the aquarium, remove your finger and lower the free end of the tube to the bottom of the container filled with yellow water. Place the other end at the bottom of the container of cold blue water. Remove your finger. This should establish a siphon that slowly drains the cold blue water into the container in the aquarium.

5. Observes what happens.

Figure 6-8. Submarine waterfalls apparatus.

Questions

1. As the blue water fills the container, what happens to the yellow?

2. Does the yellow water mix easily into the surrounding clear water? With the blue water?

3. What happens to the blue water after it fills the container? Why?

4. How rapidly does the blue water fall in the underwater waterfalls? Why?

5. Is there much mixing between the blue and clear waters? If so, where does it occur? Why?

6. After a layer of blue water has accumulated at the bottom of the aquarium, does the cascading blue water from the waterfalls generate waves in the blue layer on the bottom? Why do these form?

## Procedure B

Repeat steps 1 to 5, but substitute cold salt water for cold, freshwater.

## Questions

1. As the blue water fills the container, what happens to the yellow?


2. Does the yellow water mix easily into the surrounding clear water?
   Into the blue water?


3. What happens to the blue water after it fills the container?
   Why?


4. How rapidly does the blue water fall in the waterfalls?
   Why?


5. Is there much mixing between the blue and clear waters?
   Where does this occur?
   Why?


6. Allow a blue layer to accumulate on the bottom of the aquarium. Are waves generated
   on the layer of blue water as water cascades down from the waterfalls?
   Why?


7. Which water mass had the greater density, the cold, fresh water or cold, saline water?
   (Devise an experiment to determine this answer.)


8. Which water mass fell more rapidly?
   Why?


9. Which water mass mixed more easily with the surrounding water?
   Why?

## 6-6. VOLUME TRANSPORT

**Volume transport** is the amount of water transported by a current in a given period of time. A good approximation of volume transport can be calculated if the average current speed, thickness and width are known. Volume transport will be equal to the speed of the water times the cross section area of the current (width multiplied by thickness).

$$V.T. = speed \times thickness \times width$$

This formula only provides an approximation for several reasons. Width multiplied by thickness gives a rectangular cross sectional area for the current. It is unlikely that any naturally occurring current is rectangular in cross section. Also, speed will vary across the current. It is usually fastest in the center and declines outward. Despite its limitations, this equation still provides an easy and fast approximation of volume transport.

To determine the volume transport within the containers used in this lab, you will need current speed, thickness of the water and width of the container. Each circular current occupied the entire width of the container. In half of this width the water was flowing in one direction and in the other half it flowed in the opposite direction. Thus, we can use half of the width of the container as the current's width. The entire thickness of water in the container was flowing, but as two separate currents. The surface current was above the pycnocline and its thickness was from the surface to the pycnocline. The deep current was below the pycnocline and its thickness extended from the top of the pycnocline to the bottom. Current speeds were measured as part of Exercise 2 of this lab.

Copy the pertinent data from Exercise 2 in the blanks below and calculate the volume transport for these currents.

Width of container: _____cm   Half width of container: _____cm
Surface layer thickness: _____cm
Bottom layer thickness: _____cm

Red coloring     _____sec to travel 10 cm     _____cm/sec
Blue coloring     _____sec to travel 10 cm     _____cm/sec

*Volume Transport*
Red coloring     _____cm/hr
Blue coloring     _____cm/hr

## 6-7. SEASONAL MIGRATION OF CURRENTS

Over the course of a year, the position of the Sun changes relative to Earth's surface (Figure 6-9). At noon on March 20 or 21, the Sun is directly above the equator. This is called the **equinox**. From March 21 until June 21, the position of the Sun at noon moves farther northward each day. On June 21 or 22 at noon, the Sun is directly overhead at

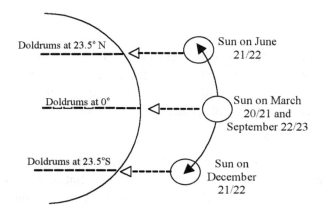

Figure 6-9.  Movement of the Sun relative to Earth's surface over the course of a year.

23.5°N (**Tropic of Cancer**), as far north as the Sun moves. This is the **summer solstice** in the Northern Hemisphere and **winter solstice** in the Southern Hemisphere. From June 21 until September 21, the position of the Sun at noon moves farther southward each day. At noon on September 22 or 23, the Sun again is directly above the equator and the second equinox of the year occurs. From September 21 until December 21, the Sun continues to move southward until, on December 21 or 22 at noon, the Sun is directly above 23.5°S (**Tropic of Capricorn**), as far south as the Sun moves. This is the winter solstice in the Northern Hemisphere and summer solstice in the Southern Hemisphere. Finally, on March 20 or 21, the Sun returns to its original location above the equator at noon.

The position of the Sun relative to Earth's surface is important because it determines the location of high and low atmospheric pressures, winds and ocean surface currents. The distribution of winds and currents illustrated in Figure 6-3 is only valid for 2 days each year, the equinox, when the Sun is directly above the equator. For the remainder of the year, the winds and currents occupy different locations.

Warm air rises in the doldrums—also called the **intertropical convergence zone**—cools and flows poleward. Some of the air reaches the poles and then descends, forming the polar highs and the polar easterlies (Figure 6-10). Much of the air, though, descends about 30° latitude, north and south of the doldrums, forming the horse latitudes, also called the **subtropical highs**. This air then flows back to the doldrums as the Trade Winds or blows poleward for about 30° latitude, where it collides with the Polar Easterlies and forms the subpolar lows. While the Trade Winds and Westerlies each generally blow across 30° of latitude wherever they are located, the polar easterlies greatly expand in winter and contract in summer. The location of the ocean gyres and countercurrents changes daily as the location of the Trade Winds and Westerlies change.

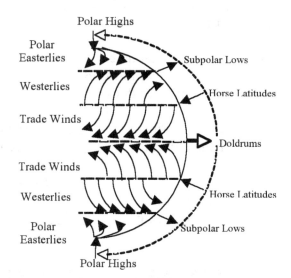

Figure 6-10.  Distribution of atmospheric high and low pressures and winds on the equinox. Arrows indicate direction of air movement.

# EXERCISE 5.  MIGRATION OF CURRENTS

Below are maps of the North and South Atlantic Ocean for March 21, June 21, September 22 and December 21. On the maps to the left, indicate the locations and latitudes where each of the following would be found for the dates indicated: doldrums, subtropical highs, and subpolar lows. On the same map, draw the Trade Winds, Westerlies and Polar Easterlies. On the maps to the right, draw the north and south gyres and their counter currents for the indicated dates.

Questions

1. Are the counter currents always on opposite sides of the equator?  Why?


2. On the equinox, the counter currents flow more swiftly than they do on the solstice. How can Coriolis explain this?


3. If the Earth rotated in the opposite direction, how would this alter the flow of the gyres and the counter currents?

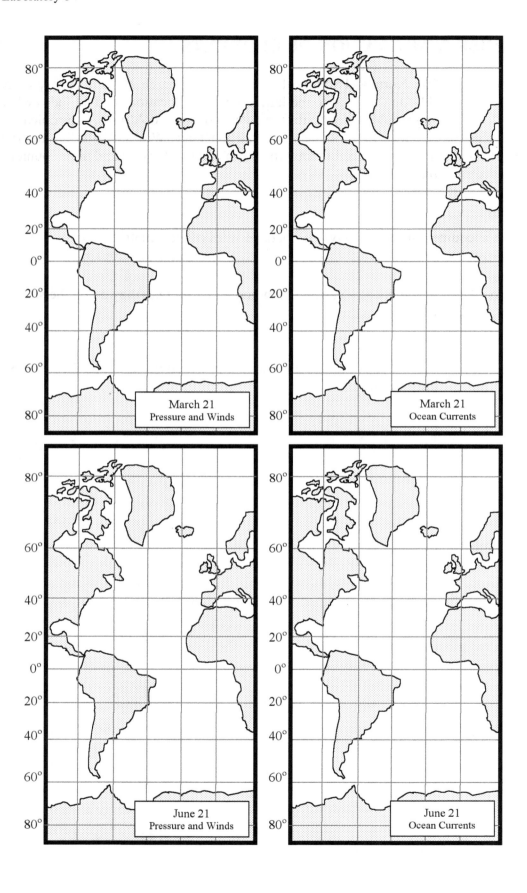

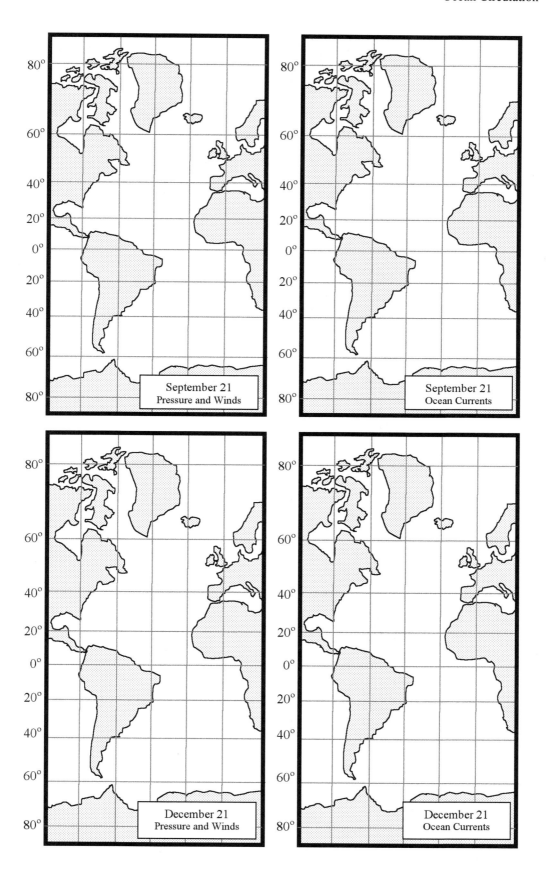

# Laboratory 7

# Waves

## 7-1. TYPES OF WAVES

An ocean **wave** is an alternating rise and fall of a portion of the water surface or of a density boundary within the water column. Waves can be produced by the wind, movement of objects into, out of or within the water, and vibrations (earthquake) of the basin containing the water body. Production of a wave represents the transfer of energy into the water from the wave-inducing source. Two basic types of waves are progressive waves and standing waves or seiches. A **progressive wave** moves forward across the water from the area in which it was formed. At any moment, it occupies only a small portion of the entire surface. **Seiches** remain where they are produced and appear as a rocking back and forth (up and down) of a large portion of the water surface or boundary layer.

## 7-2. WAVE PARAMETERS

Regardless the type, all waves possess various features that can be measured. These include (Figure 7-1) wave length, height, amplitude, period, and frequency. From these variables, wave stability (steepness), and celerity (speed) may be calculated.

**Wave crest** is the highest point on a wave and **wave trough** is the lowest. **Wave length** (**L** or $\lambda$) is the length of one complete wave form, as measured from wave crest to adjacent wave crest (or trough to trough). **Wave height** (**H**) is the vertical distance from wave crest to wave trough. **Amplitude** (**A**) is the vertical displacement of the crest or trough from the flat, undisturbed water level. It is equal to one-half the wave height (A = 1/2 H).

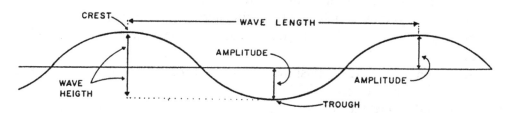

Figure 7-1. Wave parameters.

**Wave period** (**T**) for a progressive wave is the amount of time required for one wave length to pass a fixed point. For a standing wave, the period is the amount of time for one complete oscillation of the water surface. **Wave frequency** (**F**) for a progressive wave is the number of waves passing a fixed point in a unit of time and is equal to one divided by the period (F = 1/T). For a standing wave, the frequency is the number of oscillations in a unit of time. **Celerity** (**C**) only applies to progressive waves and is the velocity with which a wave form travels. Celerity is not exactly the same as speed. **Speed** implies that a mass is moving over a distance in a period of time. In progressive waves, it is not the water that moves forward as the wave travels, but the wave form. Because wave form has no mass, it would be inappropriate to use the term speed. Celerity can be measured directly or calculated by dividing the wavelength by wave period (C = $\lambda$/T). **Wave stability** or steepness is the ratio of wave height to wave length (stability = H/$\lambda$). A wave becomes unstable and collapses or **breaks** when H/$\lambda$ > 1/7.

## EXERCISE 1.  PROGRESSIVE WAVE PARAMETERS

Procedure

1. Use a long aquarium or wave tank. The longer the tank, the easier it will be to measure the wave parameters. If using a long tank, place the tank so that one end is raised and 10–20 cm of the bottom of the tank is exposed on the higher end. This exposed area will allow the waves to break and not be reflected off the end of the tank. Make all measurements about midway between the "shore" and the wave source.
2. With an erasable marker, lightly mark the standing water level along the side of the tank.
3. Using a board as a plunger, raise and lower the board in the water every 1 sec at the deep end of the tank to generate regular waves. Do not allow the board to completely emerge from the water or hit bottom.
4. As a chosen wave passes midway between the board and "shore," mark the crest and trough heights on the side of the tank. Stop generating waves. Measure the wave height from the two marks you have made and record the value below. Calculate and record the amplitude of the wave. Record water depth at this point.
5. Generate more waves as before. At exactly the same time that you mark the crest of one wave on the side of the tank, have your lab partner mark the adjacent wave crest. Cease making waves and measure and record the wavelength.
6. Generate more waves as before. Determine and record the period and celerity of the waves. Repeat this procedure three times and calculate the average period and celerity.
7. Count the number of waves past the midpoint of the tank in 1 min. Calculate the frequency for 1 sec. Record the data.
8. Calculate and record the celerity of the wave using the formula C = $\lambda$/T.
9. Calculate and record the frequency of the waves using the formula F = 1/T.

10. Repeat steps 1 through 9, but generate waves every half second for one data set and then every 2 sec for the other data set.

**Data Table**

| Parameter | 0.5-sec Wave | 1-sec Wave | 2-sec Wave |
|---|---|---|---|
| Height | | | |
| Amplitude | | | |
| Water Depth | | | |
| Length | | | |
| Period | | | |
| Period | | | |
| Period | | | |
| Average Period | | | |
| Celerity (measured) | | | |
| Frequency | | | |
| Celerity (calculated) | | | |

Compare and contrast the three wave sets. Does the period with which the waves were generated greatly effect their parameters?

## 7-3.  WAVE MOTION IN DEEP AND SHALLOW WATER FOR PROGRESSIVE WAVES

In deep water, the progressive wave form moves forward rapidly, but individual molecules of water do not. With each passing wave the water molecules travel forward and up with the crest of the wave and then backward and down with the trough. They return *almost* to their original position after the wave passes. This nearly circular path is called an **orbit**. The diameter of the orbit at the surface is equal to wave height. Orbital diameter decreases with depth (Figure 7-2) and is zero at wave base.

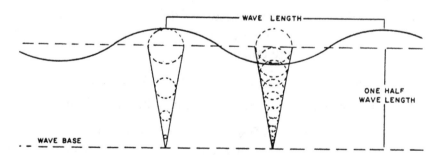

Figure 7-2. Circular orbital motion in a deep-water progressive wave.

**Wave base** is the maximum depth to which a passing wave imparts motion to the water. It is equal to one-half the wavelength (wave base = $\frac{1}{2}\lambda$). **Deep water** for a given wave is defined as water having a depth greater than $\frac{1}{2}\lambda$ for that wave. **Celerity** (C) of waves in deep water may be calculated as:

$$C = gT/2\Pi r = 1.56T = 1.25\sqrt{\lambda} \text{ (where } g = 9.8 \text{ m/sec}^2).$$

**Shallow water** for a progressive wave is defined as a water depth less than or equal to 1/2λ. In shallow water, frictional drag on the bottom slows the wave and distorts the circular orbits into ellipses (Figure 7-3). The ellipses become flatter with depth and are reduced to a simple back-and-forth motion on the bottom. In shallow water, wavelength decreases and height increases as

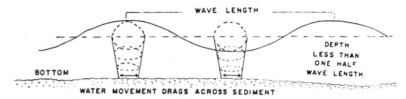

Figure 7-3. Elliptical orbital motion in a shallow-water progressive wave.

waves are "squeezed" together. Eventually, the waves become unstable and crash forward. Shallow-water wave celerity decreases as water depth decreases and celerity can be calculated as:

$$C = \sqrt{gd} = 3.1\sqrt{d}$$

(where g = 9.8 m/sec$^2$ and d is depth)

As waves break against the shore, the elliptical orbit is broken and remaining wave energy is expended transporting water across the beach face. Gravity then causes the water to flow back to the sea.

## EXERCISE 2.  CALCULATING CELERITY OF PROGRESSIVE WAVES

Procedure

1. Using the data from Exercise 1 of this laboratory, determine if the progressive waves generated were deep-water or shallow-water waves at the point where the measurements were taken.

    0.5-sec wave_____ 1-sec wave_____ 2-sec wave_____

2. Using the appropriate equations given below, calculate wave celerity in cm/sec.

    $$C = 1.25 \sqrt{\lambda} \text{ m/sec or } C = 1.56T \text{ (deep-water waves)}$$
    $$C = 3.1 \sqrt{d} \text{ m/sec (shallow-water waves)}.$$

    0.5-sec wave_____ 1-sec wave_____ 2-sec wave_____

3. Compare the results obtained mathematically with those obtained experimentally in Exercise 1. How well do they compare? Why might they differ?

# EXERCISE 3.  MOVEMENT OF WATER IN WAVES

Procedure A

1. Keep the tank tilted, as in Exercise 1. Place a small cork near the center of the tank about 20 cm from the board used to generate waves. Mark the position of the cork on the side of the tank.
2. Generate waves by raising and lowering the board every second, as in Exercise 1. Continue this for about 45 sec, or until the cork becomes stuck to the side of the tank. In the space below, record the number of waves generated and measure the distance the cork travels.

Questions

1. Number of waves: _____

2. Distance cork travels: _____cm

3. Describe the motion of the cork in the water.

4. Distance cork moves forward with each wave: _____ cm

5. If particles in waves move in circular orbits, why does the cork gradually move forward?

Procedure B

1. Place the cork in the center of the tank about 10 cm from the "shore."
2. Generate waves as in Procedure A. Continue waves until cork is stranded on the "shore." Start again if the cork becomes stuck on the side of the tank. In the space belows record the number of waves and measure the distance the cork travel.

Questions

1. Describe the motion of the cork in the water.

2. Number of waves: _____

3. Distance cork travels: _____ cm

4. Distance cork moves forward with each wave: _____ cm

5. How does the distance traveled per wave compare to what was determined in Procedure A? Why is there a difference?

## 7-4. PROGRESSIVE WAVE REFLECTION AND INTERFERENCE

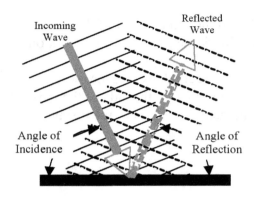

Figure 7-4.  Reflection of progressive waves as viewed from above.

**Wave reflection** occurs when a wave rebounds off a steep shore or vertical wall. Wave energy is transformed into a new wave traveling in the opposite direction from which the original wave came (Figure 7-4). The angle at which the wave strikes (**angle of incidence**) will be equal to the angle at which the new wave departs (**angle of reflection**). If the incoming wave strikes head on, the reflected wave will travel outward in the opposite direction (at 180°).

Waves interact as they pass through each other. **Constructive interference** (Figure 7-5A) occurs when like parts of a wave coincide. The maximum constructive interference is when crests overlap or troughs overlap. The resulting composite wave will be almost equal in size to adding the two parts of the waves together. For example, if one wave height is 2 ms and the other wave height is 3 ms, the composite wave will have a wave height of almost 5 ms at the moment the two crests and troughs overlap. **Destructive interference** (Figure 7-5B) occurs when dissimilar parts of waves coincide. The resulting composite wave will be almost equal in height to subtracting the two parts of the waves. For example, if both waves have a wave height of 3 ms, the water surface will remain flat as a trough and crest of thetwo waves momentarily coincide.

As waves pass through each other, both constructive and destructive interference occur as various parts of the waves interact. Interference does not permanently alter a wave's properties. After momentarily interacting, each wave continues onward as though no interaction has occurred. Interference happens whenever two or more waves occupy the same space, regardless of the direction in which the waves are traveling or their celerity.

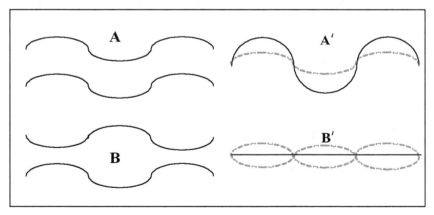

Figure 7-5. A and A'. Constructive wave interference. A shows waves before interacting. A' shows waves overlapping (dashed line) and the resulting wave in black. B and B'. Destructive wave interference. B shows waves before interacting. B' shows waves overlapping (dashed line) and the resulting sea surface in black.

## EXERCISE 4. PROGRESSIVE WAVE REFLECTION AND INTERFERENCE

Procedure

1. Place the tank so that it is level. Allow the water to calm.
2. From one end of the tank, generate a single large wave with a board. Immediately remove the board from the water.
3. Observe what happens as the wave strikes the opposite end of the tank. Allow the surface to calm.
4. Generate a new wave and record how many times the wave travels back and forth along the tank.
   Number of times traveled along tank: _____
   Does reflection appear to consume much of the wave's energy? Why?


5. Generate three waves. How many waves are reflected? _____
6. Simultaneously generate waves at opposite ends of the tank. Observe what occurs as the waves pass through each other.
   How was constructive interference expressed? Destructive interference?


## 7-5. WIND-GENERATED PROGRESSIVE WAVES

**Surface waves** form on the surface of the ocean at the air-water interface. Most surface waves are produced by the wind. As air moves across the water surface, **capillary waves** are the first to form. They are characterized as having a wavelength of less than 1.7 cm, period less than 0.1 sec, rounded crests and V-shaped troughs. Capillary waves radiate outward in regular arcs of long radius and provide an irregular, rough surface upon which the wind may act.

As the wind blows across a wave, it pushes against the wave on the **windward side** (side toward the wind) and creates an area of low pressure on the **leeward side** (downwind) of the wave (Figure 7-6). The low pressure pulls the wave forward as the wind pushes the wave from behind. With continued wind, capillary waves can grow to become ripples, short choppy waves, and eventually **white caps**. For wind-produced waves, wavelength and wave height increase with increasing wind velocity, duration, and **fetch** (area over which the wind blows).

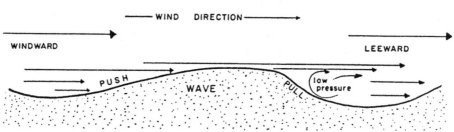

Figure 7-6. Wind transferring energy to waves.

## EXERCISE 5.  WIND-GENERATED PROGRESSIVE WAVES

Procedure

1. In a large tub or small swimming pool, construct a sloping beach that is about 7 cm thick and 15 cm wide. Submerged sediment should continue for about 15 cm into the water. (Setup is to be reused in Exercise 6.)
2. Allow all currents and waves to stop. From the side opposite the beach, blow very hard across the water toward the beach for 4 sec. (A fan with different speed settings may be substituted for blowing.) Observe what occurs. Allow the waves to cease. Blow gently across the water toward the beach for 4 sec and observe what occurs.
3. Briefly describe, in terms of relative wave size and celerity, the differences between the two wave groups.


   Did the differences result from variation in fetch, velocity or duration of the wind?


4. Allow all currents and waves to stop. Blow very hard across the water toward the beach for 10 sec. Observe what occurs. Allow the waves to cease. Blow very hard across the water toward the beach for 4 sec and observe what occurs.
5. Compare these waves with those generated in step 2 in terms of relative wave size and celerity.


   Did the differences result from variation in fetch, velocity, or duration of the wind?


## 7-6.  PROGRESSIVE WAVES AND SEDIMENT

Throughout most of the ocean, the majority of progressive waves are deep-water waves. They exert no influence on bottom sediments because their wavelength is small compared to the water's depth. Only in shallow water, where waves "touch bottom," and on the beach, where waves break, do waves exert an influence on bottom sediment. The back-and-forth motion of the water caused by the waves produces turbulence that maintains small particles in suspension and allows them to be transported. Size of particles in suspension and depth to wave base vary with wave size. In areas exposed to continuous wave action, especially the beach, smaller particles are eventually transported to less turbulent areas, such as deeper water or protected bays, and are deposited.

# EXERCISE 6.  PROGRESSIVE WAVES AND SEDIMENT

Procedure

1. Using a board, generate several medium size waves opposite the beach constructed in Exercise 5. Describe what occurs to the sediment 4, 8, 10, and 30 cm seaward from the shore.

2. Generate several smaller waves and describe what occurs to the sediment in the same locations as above.

3. From these observations, how important would you expect waves to be in distributing sediment on the ocean floor?  Why?

4. Which wave group contained more energy?  Why?

5. What does the motion of the sediment tell about wave base for the two wave groups?

# 7-7.  PROGRESSIVE WAVE REFRACTION

**Wave refraction** is the bending of waves into an area where they travel slower. Refraction first occurs as a progressive wave passes from deep water into shallow water (Figure 7-7). Upon entering shallow water, friction with the bottom slows the wave and the wave gradually bends towards the beach. Because shallow-water wave celerity decreases with water depth, refraction increases shoreward as water depth declines. The wave will be *nearly* parallel to the beach as it breaks. Regardless of the direction of approach or irregularity of the coast, waves break nearly parallel to the shore.

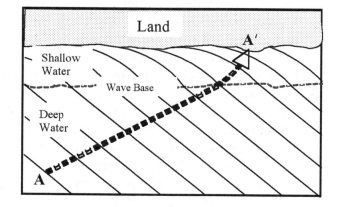

Figure 7-7. Refraction of progressive wave. A to A′ maps movement of a point on the wave crest as it approaches shore. Refraction begins at wave base and bends the wave towards the shoreline. Diagonal lines are wave crests.

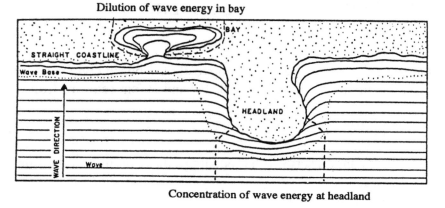

Figure 7-8. Refraction of waves along an irregular coastline.

Wave refraction strongly influences erosion and deposition. A **headland** is land that extends out into the sea as a peninsula. Waves refract towards headlands, thereby concentrating wave energy upon them, subjecting them to higher waves and greater wave-related erosional forces (Figure 7-8). In contrast, refraction bends waves outward into bays or coves, diluting the wave energy as the wave is stretched along the entire shore (Figure 7-7). Waves are smaller and sediment may be deposited as a wave enters a bay and begins to refract.

## EXERCISE 7.   WAVE REFRACTION, EROSION AND DEPOSITION

Procedure

1. In a large tub or small pool, construct sandy beaches to reproduce each of the diagrams (A–D) illustrated on the following page. Features above sea level are stippled and outlined by a solid line. Significant shallow submerged features are surrounded by dashed lines and labeled "submerged bar."
2. Using a board, produce several small waves directed toward the beach area from the end of the tank opposite the beach. Choose a single wave and, on the diagram entitled "Small Direct Waves," draw a series of lines, similar to those in Figure 7-7, showing the position of the wave crest as the wave approaches the shore.
3. Produce several small waves at an angle of about 45° to the beach. Choose a single wave and, on the diagram entitled "Small Oblique Waves," draw a series of lines showing the position of the wave crest as the wave approaches the shore.
4. If continued long enough, the small waves would eventually erode some areas and deposit sediment in others. Because of time restrictions, we shall obtain similar results using medium to large waves. Produce several larger waves. Continue to generate these waves until a recognizable change is produced in the distribution of the land and sea along the shore.
5. In the box entitled "Large Waves," draw the new distribution of land and sea over the dotted line marking the previous shoreline. Label areas of deposition and erosion.

## Small Direct Waves            Small Oblique Waves            Large Waves

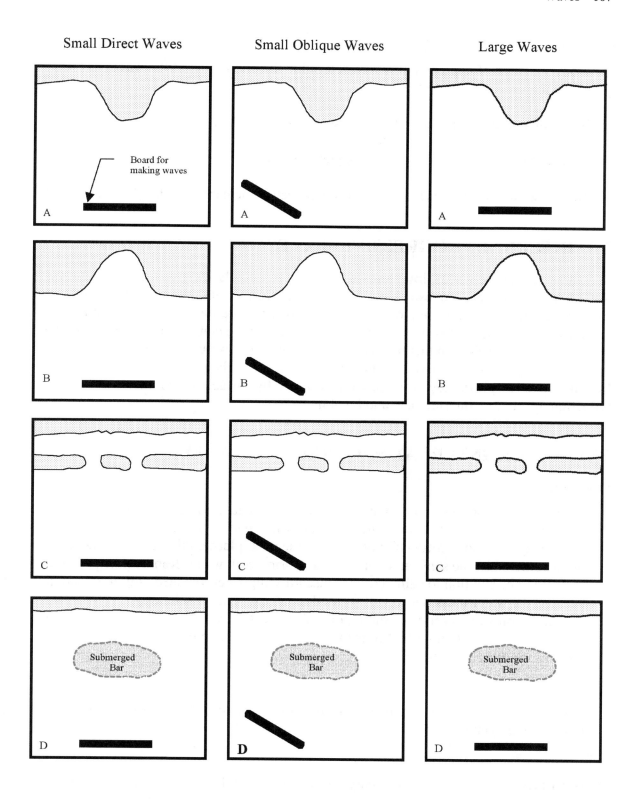

Questions

1. Does the angle of approach of waves determine the orientation of the waves as they strike the beach?

2. What areas were subjected to the most severe erosion? Why?

3. What areas were subjected to major deposition? Why?

4. Did the waves being funneled into a narrowing valley increase wave height in model B?

## 7-8.  INTERNAL PROGRESSIVE WAVES

In contrast to surface waves, **internal waves** occur on boundaries (pycnoclines) between water masses of different densities within the ocean. Surface and internal waves are similar, but internal waves tend to be smoother, slower, more stable, and can attain greater wave heights than surface waves. This is because the density difference between the water layers is less than that between air and water. The water surrounding the internal wave provides support for higher waves, but slows the progress of the wave. Internal waves display many of the properties shown by surface waves, such as reflection, refraction, interference, and breaking.

## EXERCISE 8.  INTERNAL PROGRESSIVE WAVES

Procedure

1. Follow the instructions in Laboratory 5 and arrange a tank for a water density experiment. Place the tank on a sheet of plain white paper. Use as little clay as possible to seal the partition. On one side of the tank, place cold, blue salt water and on the other side, the same amount of clear, warm, fresh water. Remove the partition and allow the solution to calm. (Do not discard setup when finished with this section because the water masses also will be used in the exercise on seiches.)

2. Forcefully blow, in a series of sharp puffs, directly downward on the water surface in the central region of either half of the tank. Observe through the side of the tank. On which surfaces do waves form?


   How do the sets of waves on the surface differ in form (size, height, speed, etc.) from those on the pycnocline?


3. Blow across the surface so that sharp puffs of air strike the central region of either half of the tank at about 45°. How do the surface waves differ from the internal waves?

# 7-9. STANDING WAVES

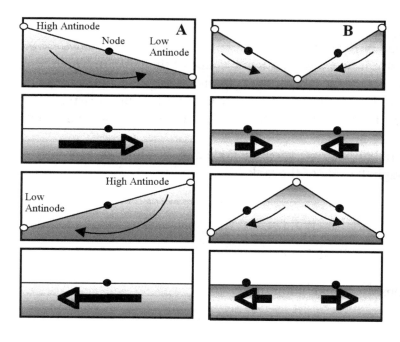

Figure 7-9. A. Simple seiche. B. Complex seiche. Arrows indicate direction of flow and velocity increases with arrow thickness. Black circles are nodes and white circles antinodes.

**Standing waves** (also called **stationary waves** or **seiches**) appear as a rocking back and forth (up and down) of the water surface or an internal boundary within the water column (Figure 7-9). The line about which the seiche rocks is called a **node** and it experiences no vertical motion. In contrast, the **antinode** is where the vertical motion is the maximum. A simple seiche has only one node in the center of the basin and two antinodes at the opposite ends. Complex seiches may have more nodes and antinodes (Figure 7-9B). The **period** of the seiche is the time required for the water surface to rock back and forth once and is dependent on the basin length and water depth. For a **closed basin**, similar to a container, the period is $T = 2l\sqrt{gd}$, where l is the basin length along which the seiche oscillates, d is water depth and g is the 980 cm/sec$^2$. For an **open-ended basin**, such as a bay connected to the ocean, the period is $T = 4l\sqrt{gd}$. The node for an open basin is always at the connection to the sea.

Seiches display interference, refraction, and reflection, but unlike progressive waves, they do not move water in circular to elliptical orbits. In a seiche, water flows forward until the surface is inclined between the antinodes, stops and then flows in the reverse direction. Maximum velocity is achieved when the surface is horizontal. No flow occurs when the surface is at a maximum tilt (antinode).

A **rotary seiche** (Figure 7-10) is a special type of seiche in which the position of the antinodes rotates about the basin as the surface rocks back and forth. The node is reduced to a point about which the surface revolves. The antinodes inscribe circles as they migrate around the basin.

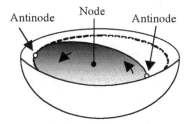

Figure 7-10 Rotary seiche. Black dot is the node. White dots are antinodes. Dashed line indicates movement of antinode around basin. Arrows show direction of motion.

## EXERCISE 9.  SEICHES

Procedure A

1. Use the container as arranged in Exercise 7.
2. Development of seiches is strongly dependent on the shape of the basin in which the water body is contained. Raise one side of the container along its length about an inch and then lower it rapidly. (Do not drop it!) Try not to create progressive waves. View the oscillation of the water from the side. Locate the node and antinodes on the surface.
3. Add a drop of coloring about half way between the node and the antinode and observe its movement through the side of the tank.

Questions

1. Do the surface and pycnocline oscillate in unison? _____

2. Does the oscillatory relationship change with time? _____
   If so, how?

3. Does one continue to oscillate longer than the other? _____
   Which one and why?

4. What is the period of the seiche at the surface? _____ Internal layer?_____

5. Why is the period of an open-ended basin twice that of the period of a closed basin?

6. Describe the motion of the drop of coloring added to the water after the seiche had been established.

Procedure B

1. Half fill a round white bowl with pale blue water.
2. Place the bowl on a table and push it forward to establish a standing wave.
3. Measure the period of the wave after all progressive waves have ceased.
   Period: _____
   Do the antinodes stay in one location as the surface oscillates or do they revolve around the bowl? _____ Why?

4. Allow the water to calm. Hold the bowl on the table with both hands and start moving the bowl in a circular motion for a few seconds. Follow the movement of the water in the bowl. Add a drop of red coloring to the water and describe its motion.

How long does it take for the inclined water surface to move around the bowl?
_____

Is this a rotary seiche? _____ Why?

Does the water surface rock up and down or is it just an inclined surface revolving around the bowl? _____

5. Establish a seiche as in step 2. Then hold the bowl on the table with both hands and start moving the bowl in a circular motion.
How does the water surface behave differently than was observed in step 4? Why?

# Laboratory 8

# Tides

## 8-1. ORIGIN OF TIDES

**Tides** are bulges of water produced by the **gravitational attraction** of the Sun and Moon on the ocean and by the **centrifugal effect** resulting from rotation of the Earth-Sun and Earth-Moon systems. **Gravity** is the force of attraction between two bodies. It is proportional to the mass of the objects and inversely proportional to the square of the distance between them. The greater the mass and smaller the distance, the stronger is the gravitational attraction.

$$G \propto M_1 M_2 / d^2$$

where G is gravity, $M_1$ and $M_2$ represent the mass of body 1 and body 2 and d is distance between the center of the two masses.

All objects in space exert a gravitational attraction on the oceans, but only the Sun, because of its great mass, and the Moon, because of its proximity, exert sufficient attraction to produce a noticeable bulge of water. Despite the Sun's great mass (332,960 × the mass of the Earth), the distance between the Earth and Sun (149,700,000 km) reduces the solar tidal attraction to only 46% of the attraction exerted by the Moon. The Moon is much smaller than the Sun and even smaller than the Earth, but it is closer to the Earth and thus exerts a gravitational attraction disproportionately strong for its size.

**Centrifugal effect** is directed away from the center of rotation as a body moves in a curved path. As the Earth revolves on its axis, centrifugal effect attempts to throw objects from the Earth's surface into space. If Earth's gravity was sufficiently weak or its speed of rotation sufficiently great, objects would be hurled into space. Centrifugal effect keeps water in a bucket when it is spun in a vertical circle. At the top of the circle, centrifugal effect pushes the water against the bottom of the bucket and stops it from falling out.

As the Earth travels in its orbit about the Sun, centrifugal effect acts on the side of the Earth away from the Sun and attempts to fling the ocean into space. Gravitational attraction between the Earth and ocean is greater than centrifugal effect and the ocean remains on the Earth, but a bulge is produced in the ocean (Figure 8-1).

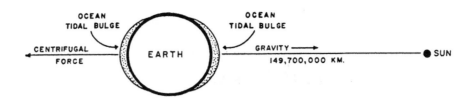

Figure 8-1.  Sun-induced tidal bulges.

In contrast to common belief, the Moon does not revolve around the Earth, but rather the Earth and Moon rotate about a point called the **barycenter** located inside the Earth about 4700 km toward the Moon from Earth's center (Figure 8-2). As the Earth and Moon revolve around this point, centrifugal effect produces a tidal bulge on the side of the Earth facing away from the

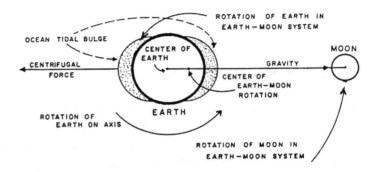

Figure 8-2.  Moon-induced tidal bulges.

Moon. Thus, the Sun and Moon each produce two tidal bulges in the ocean. One pair of bulges is produced by gravitational attraction and faces the Sun and Moon. The other pair of bulges is generated by centrifugal effect and faces away from the Sun and Moon.

## 8-2.  TIDES AND TIDAL RANGES

As the Earth revolves on its axis, coastal areas pass through the tidal bulges and alternately experience high and low tides. As the crest of a bulge passes, **high tide** occurs. Troughs between bulges result in **low tides**. **Tidal range** is the difference in height between high and low tides. A continuous record of the daily change in sea level because of tides is a **tidal curve** (Figure 8-3).

Tidal curves show that tides change daily in range and time of occurrence. The variation

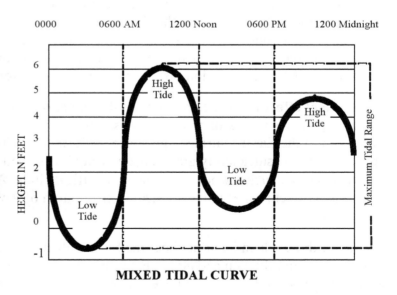

Figure 8-3.  Tidal curve showing mixed tide and maximum tidal range.

results from the complex interaction of the Earth, Sun, and Moon. The interaction of these three bodies is so complex that a given coast may experience the same tidal conditions only once each 19 yrs.

## A.   SPRING   TIDE

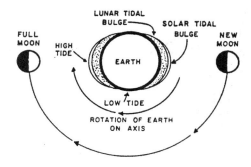

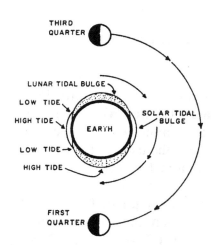

## B.   NEAP   TIDE

Figure 8-4.  Spring and neap tides. Spring tides occur with the new and full Moon and neap tides with the first and last quarter Moons.

**Solar tides**, the tides produced by the Sun, occur approximately at the same time each day because the Earth rotates on its axis about every 24 hr. Coastal areas experience a solar tide every 12 hr. **Lunar tides** are more complex because the Moon moves forward in its orbit as the Earth rotates on its axis. Thus, a point on the Earth directly below the Moon requires 24 hr and 50 min before it is again below the Moon. Lunar tides occur 50 min later each day. Coastal areas experience a lunar tide every 12 hr and 25 min.

## 8-3.  SPECIAL MONTHLY TIDES AND PHASES OF THE MOON

About twice each month, lunar and solar tidal bulges are in phase and experience constructive interference. Solar and lunar tidal bulges overlap, as do the troughs, and produce a very high high-tide and a very low low-tide. Bimonthly maximum tidal ranges are called **spring tides**. They have nothing to do with the season of spring. Spring tides result from the "straight" line arrangement of the Earth, Moon and Sun (Figure 8-4) and occur during **full** and **new Moons**. Similarly, twice each month lunar and solar tidal bulges are out of phase and experience destructive interference. Solar tidal bulges overlap lunar troughs and lunar tidal bulges overlap solar troughs. Minimal tidal ranges result. These tides are called **neap tides**. They occur during the **first** and **third quarter Moon**, when the alignment of the Moon, Earth and Sun form a 90° angle (Figure 8-4).

Terms describing the **phases of the Moon** can be confusing (Figure 8-5). The Moon generates no illumination of its own, but reflects sunlight. Only that portion of the Moon facing the Sun that can be seen from Earth appears illuminated in the sky. **New Moon** occurs when the Moon lies directly between the Earth and the Sun. Because the illuminated side of the Moon faces away from the Earth, no Moon is visible in the sky.

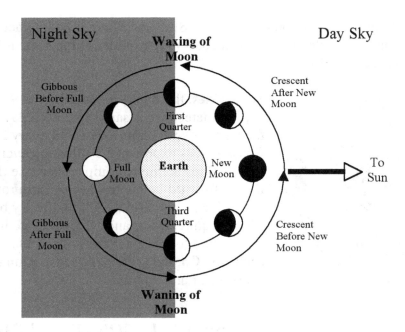

Figure 8-5. Phases of the Moon. Light part of lunar face indicates portion seen from Earth.

For the next 7 days, the illuminated face of the Moon gradually rotates towards Earth and the lunar crescent expands. This is called the **crescent after the new Moon**. Crescent is always less than half. By the 7th day, half of the Moon is illuminated in the evening sky and this is called the first quarter Moon. For the next 7 days, the face of the Moon continues to expand and more than half of the Moon is illuminated. This is called **gibbous before the full Moon**. Gibbous means greater than half. The 14th night is the full Moon. For the next 7 days, the illuminated face of the Moon gradually turns away from Earth as the Moon continues in its orbit. Each night, less of the Moon is illuminated, but always more than half. This is called **gibbous after the full Moon**. On the 21st night, half of the Moon is illuminated and this is the third-quarter (last quarter) Moon. For the next 7 days, the Moon becomes a shrinking crescent. This is called the **crescent before the new Moon**. On the 28th day, the Moon again becomes a new Moon.

The monthly trip of the Moon "around" the Earth takes 28 days and is divided into quarters, beginning with the new Moon. The terms **first** and **last quarter Moon** refer to positions in the orbit, not the proportion of the illuminated lunar disk visible from Earth. Half of the Moon is illuminated, at the first and last quarter Moon. From new Moon to half Moon is called the **evening Moon** because the Moon appears in the sky after noon and before midnight. From full Moon to new Moon is called **morning Moon** because the Moon appears after midnight and before noon. **Waxing of the Moon** describes the period between new Moon and full Moon, as the illuminated lunar disk expands. **Waning of the Moon** refers to the period between the full Moon and new Moon. Each night, less of the Moon is illuminated.

## 8-4.  OTHER CONTROLS ON TIDAL RANGE

In addition to the alignment of the Sun and the Moon, tidal range is also controlled by variation in gravitational attraction, declination of the Earth relative to the orbits of the Sun and Moon, and distribution of land. The orbit of the Earth around the Sun, and the Moon and Earth about each other, are elliptical, not circular. At times, the Earth and Sun and Earth and Moon are closer together (Figure 8-6). Because gravity decreases as objects move apart and increases as they move together, the tidal bulges respond to changes in the orbits. When in **perigee**, the Sun and Earth or Moon and Earth are closest together in their orbit and tidal bulges are higher. In contrast, when in **apogee**, the Earth and Sun or Moon and Earth are farthest apart in their orbit and tidal bulges are smaller. It is estimated that lunar tides at perigee have about a 20% greater range than at apogee.

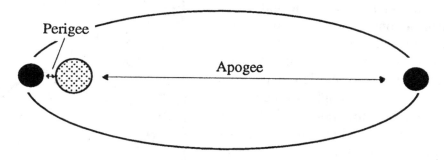

Figure 8-6.  Eliptical orbit illustrating apogee and perigee.

**Declination** is the inclination of the Earth's axis relative to the orbital plane of the Earth around the Sun or to the orbital plane of the Moon and Earth around each other. It is expressed by the maximum poleward position north or south of the equator that the Sun and Moon appear in the sky. Each month the position of the Moon varies from 28.5°N to 28.5°S. Similarly, each year the position of the Sun varies from 23.5°N to 23.5°S (Figure 8-7). The more similar latitudinal position of the Sun and Moon in either hemisphere, the greater the overlap of the tidal bulges and the greater the tidal range.

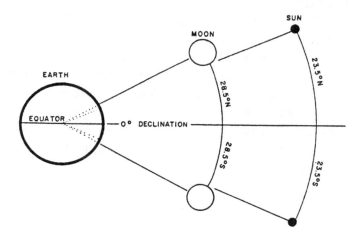

Figure 8-7.  Declination of Sun and Moon.

Distribution of land greatly influences the expression of tides. If Earth was an ocean-covered planet, tides would sweep westward unimpeded. All points along the same line of longitude would experience tides at the same time. Except around Antarctica and in the Arctic Sea, continents interrupt tidal flow by dividing the ocean into a series of interconnected basins. Basin shape and depth determine the period of the basin and how the body of water responds to tides. **Period of a basin** is the time required for a standing wave in the basin to complete one oscillation. If the period of the basin is in phase with the tidal period, the tides will reinforce the standing wave in the basin and the standing wave will rock higher with each tidal cycle.

For a large somewhat circular basin, such as the North Atlantic, the standing wave migrates around the basin as a **rotary seiche** and is called an **amphidromic system** (Figure 8-8). In this system, the ocean surface displays the same motion as would the water surface in a round bowl, as the bowl is swirled in a circular motion. The center of the amphidromic system is called the **amphidromic point**. It is a **node** and experiences no tide. Tidal range increases outward and lines connecting points of equal tidal range form concentric circles, called **corange circles**, about the amphidromic point. Lines connecting points that experience the high or low tide at the same time are called **cotidal lines**. These lines radiate out from the amphidromic point.

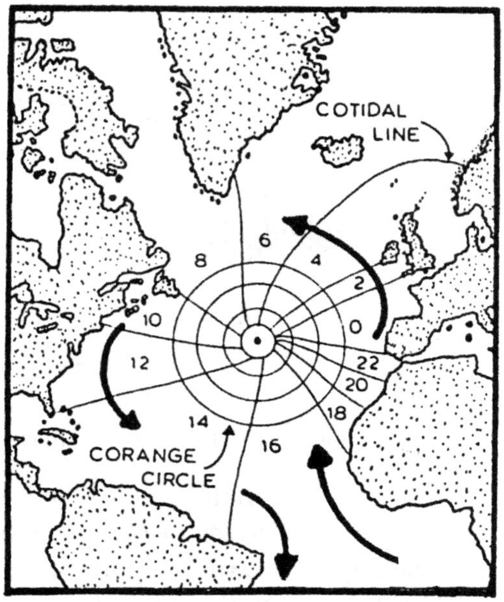

Figure 8-8. Simplified amphidromic system in North Atlantic Ocean.

In a basin with a period of about 12 hr (e.g., Atlantic Basin), **semidiurnal** (or semidaily) tides will develop. Each day two tides of about equal range will affect the coast (Figure 8-9A). **Diurnal** (or daily) tides are produced in basins with a period of about 24 hr. These basins (e.g., Gulf of Mexico) experience only one tide a day (Figure 8-9B). Basins with periods between 12 and 24 hr will display **mixed tides** and have two tides of unequal range (Figure 8-9C). The difference between the height of successive high tides or successive low tides within a day is the **daily inequality** of a mixed tide. For mixed tides, the two crests are referred to as **high high-water** and **low high-water**; the troughs are **high low-water** and **low low-water** (Figure 8-9C). Mixed tides are developed in many areas, for example, the Caribbean Sea and Pacific and Indian Oceans.

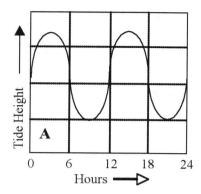 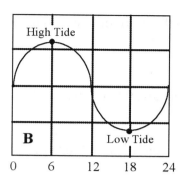 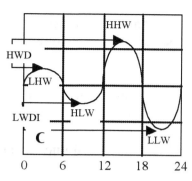

Figure 8-9. Main types of tidal curves: (A) semidiurnal, (B) diurnal and (C) mixed tides. HHW = High high water, LHW = Low high water, LLW = Low low water, HLW = High low water, HWDI = High water diurnal inequalities and LWDI = Low water diurnal inequalities for the mixed tide.

Distribution of land may alter the tidal range. Where land funnels the tide into a narrowing valley or long arm of the sea, tidal range can be greatly increased at the distal end. In contrast, tidal range will be considerably decreased where an opening restricts the inflow of the tide. For example, because tides must flow through the Straits of Gibraltar, much of the Mediterranean Sea experiences a diurnal tide range of only 0.6 m.

## EXERCISE 1.  GENERAL QUESTIONS ABOUT TIDES

1. Venus produces a single, small tidal bulge in the ocean when Earth and Venus are near each other. What force produces this tidal bulge? Why is one bulge generated and not two?

2. For each of the phases of the Moon listed below, indicate if the tidal range is:

        A.  increasing                       D.  increasing and then decreasing
        B.  decreasing                       E.  at a maximum
        C.  decreasing and then increasing     F.  at a minimum

   i.  Waxing of the Moon _____           vi.  Gibbous after the full Moon _____
   ii.  Crescent before the new Moon _____    vii.  Morning Moons _____
   iii.  New Moon _____                viii.  Waning of the Moon _____
   iv.  Gibbous before the full Moon _____    ix.  Crescent after the new Moon
   v.  Last quarter Moon _____          x.  Full Moon _____

3. Would one expect to find a tide in the Arctic Sea? Why?

4. Guantanamo Bay, Cuba, has a spring tidal range of 1.5 m when both Sun and Moon are at apogee and have a declination of 20°N. Determine approximately what part of this tidal range is produced by the Moon and what part by the Sun.

     Moon = _____    Sun = _____

5. Do amphidromic systems in the Northern and Southern Hemispheres rotate in the same or opposite directions? Why?

6. Why does tidal range increase away from the amphidromic point?

7. Why do cotidal lines radiate from the amphidromic point?

8. Assume that the Moon revolved "around the Earth" in the opposite direction than it does now. How would this alter the times of high and low tides?

9. According to Figure 8-8, the tidal seiche rotates counterclockwise in the North Atlantic, but in Chapter 6 the gyre is illustrated rotating clockwise. Why do tides and gyres rotate in different directions?

10. Do seiches in the North Atlantic and South Atlantic rotate in the same directions? Why?

## EXERCISE 2.  MEASURING THE PERIOD OF A CLOSED BASIN

Period of a basin is the time required for a standing wave in the basin to complete one oscillation. Regardless of its dimensions, the period of a basin can be calculated if the basin length (l, direction along which the wave oscillates) and average water depth (d) are known. For a rectangular basin closed to the sea, the period (T) may be calculated as follows: $T = 2l/\sqrt{gd}$, where all units are metric and $g = 980$ cm/sec$^2$.

Procedure:

1.  Measure and record the length and width of a rectangular basin.

    Length: _____ cm      Width: _____ cm

2.  Add a few centimeters of water. On the table below, record depth as d. Allow the surface to calm. Slowly raise one end of the basin about as high as the water is deep and lower it fast enough so that a seiche is established. Avoid producing progressive waves, but if they form, allow them to stop before measuring the period of the basin.

3.  At one end of the basin, measure the time needed for 6 to 10 complete oscillations of the seiche. An oscillation is a complete cycle from crest (high antinode) to trough (low antinode) to crest (high antinode). Record in the spaces below the number of oscillations and time required. Calculate and record the period of the basin using the time per oscillation measured.

4. Using the formula $T = 2l/\sqrt{gd}$, calculate and record the period of the basin.

5. Repeat steps 1 to 4, but measure the period across the width of the basin instead of along its length.

6. Raise the water level a few centimeters and record as depth B. Repeat steps 1 to 5.

7. Raise the water level again, record as depth C and repeat instructions 1 to 5.

**Period along length**

| | d (cm) | No. waves counted | Time measured | T | Calculated T |
|---|---|---|---|---|---|
| A | ____ | ____ | ____ | ____ | ____ |
| B | ____ | ____ | ____ | ____ | ____ |
| C | ____ | ____ | ____ | ____ | ____ |

**Period across width**

| | d (cm) | No. waves counted | Time measured | T | Calculated T |
|---|---|---|---|---|---|
| A | ____ | ____ | ____ | ____ | ____ |
| B | ____ | ____ | ____ | ____ | ____ |
| C | ____ | ____ | ____ | ____ | ____ |

Questions

1. How well do calculated periods correspond to measured periods?

2. How does increasing water depth alter the period? Why?

3. How does the period along the length of the basin compare to that along the width? Why?

4. If both seiches along length and across width are in the same basin, why are their periods different?

# EXERCISE 3.  AMPLIFICATION OF A SEICHE

If the period of a basin is the same as the tidal period, the tide can generate a seiche in the basin and create a much larger than normal tidal range. This explains why some bays, such as the Bay of Fundy in Canada, have an extraordinarily large tidal range. If high tide coincides with the seiche's oscillation toward shore, the tide is superimposed on the high water created by the seiche. If low tide coincides with the oscillation of the seiche seaward, the combined effect will produce very low water. We can simulate this with a closed basin.

Procedure A

1. Partially fill a basin with water.
2. Establish a seiche in the basin.
3. Select one end of the basin. Each time the water reaches its maximum height (high antinode) at the selected end of the basin, slightly raise that end as the water rocks away and then lower it as the water level rocks back.

   Each time you raise the end of the basin, you supply additional energy to the seiche because the water is flowing down a steeper slope. The water surface should rock higher with the next oscillation.

If the period of the basin is different than the tidal period, tidal range can be greatly reduced because of destructive interference. As the seiche is withdrawing water from the basin and lowering sea level, high tide occurs, which raises the water level. Then, as low tide approaches and sea level begins to decline, the seiche oscillates back toward the land, raising the water level.

Procedure B

1. Using the same basin, establish a seiche.
2. Select one end of the basin. Each time the water reaches its minimum height at the selected end of the basin and begins to flow back towards that end, raise the end slightly. As the water reaches its maximum height and begins to flow away, lower the end.

   Each time you raise the end of the basin, you create a steeper slope the wave must climb. This depletes the wave's energy. By lowering the end of the basin as the water flows away, you decrease the slope down which the water flows and lessen its speed. At some point the seiche will cease.


## EXERCISE 4.  PERIOD OF AN OPEN-ENDED BASIN

An extension of the ocean into or between land masses may act as an open-ended basin if the entrance is unrestricted. Examples of these are the Bay of Fundy, Gulf of California and Gulf of Mexico. Standing waves in open-ended basins have the node located at the mouth of the basin for a simple seiche. The period of an open-ended basin may be calculated using the following equation: $T = 4l/\sqrt{gd}$, where l, g, d and T represent the same variables as for a closed basin.

Using the data below, calculate the periods of the basins and the most probable type of tide developed for each location. If the period (T) is near 24 hr, the area would experience a diurnal tide; if near 12 hr, a semidiurnal tide; and if between 12 and 24 hr, a mixed tide.

[1 km = 1000 m; 60 sec = 1 min; 60 min = 1 hr; 24 hr = 1 day; g = 9.8 m/sec$^2$]

| Location | Avg. d(m) | L(km) | T | Type Tide |
|----------|-----------|-------|---|-----------|
| Gulf of California | 750 | 1100 | ___ | _____ |
| Bay of Fundy | 80 | 310 | ___ | _____ |
| Gulf of Mexico | | | | |
|    La.-Miss.-Ala. Coast | 177 | 900 | ___ | _____ |
|    Texas Coast | 1557 | 1800 | ___ | _____ |
|    Yucatan Coast | 430 | 1400 | ___ | _____ |

# EXERCISE 5.  INTERPRETATION OF TIDAL CURVES

1. Figure 8-10 is a tidal curve for a 1-month period. Determine the following:
   a. Type of tide. _____
   b. Spring tides: Date _____ Range _____; Date _____ Range _____ ft.
   c. Neap tides: Date _____ Range _____; Date _____ Range _____ ft.
   d. Daily inequality of high tides on the 15th: _____ ft.
   e. Daily inequality of low tides on the 7th: _____ ft.
   f. Date and height of highest high tide: Date _____ Height _____ft.
   g. Date and height of lowest low tide: Date _____ Height _____ft.
   h. Does the highest high tide and lowest low tide occur on the same day as the spring tide? Why?

2. Figure 8-11 is an enlargement of the Black Cove tidal curve for April 24, 2007. The curve shown is correct only for Black Cove and must be adjusted in time and tide height for other areas of LeDoux Sound. Correction values for nearby locations are usually published with the tidal curve of a major population center. Values for Yoakam Point, Paisley, Cyrus Point, and Strait are given below. Using these values, adjust and plot on the same graph the tidal curve for any two localities. Subtract negative time corrections or add positive corrections to the time of high or low tide given for Black Cove. Similarly, subtract negative height correction values or add positive values to the tide height indicated. Label the tidal curves as to location.

| | Time Correction | | Height Correction (feet) | |
|---------|-----------|----------|-----------|----------|
| Locality | High Tide | Low Tide | High Tide | Low Tide |
| Yoakam | −0 07 | +0 04 | −1.0 | −0.1 |
| Paisley | +0 07 | +0 06 | +0.5 | 0.0 |
| Cyrus Point | +0 34 | +0 52 | +3.3 | +0.3 |
| Strait | +1 12 | +1 54 | +2.8 | −0.2 |

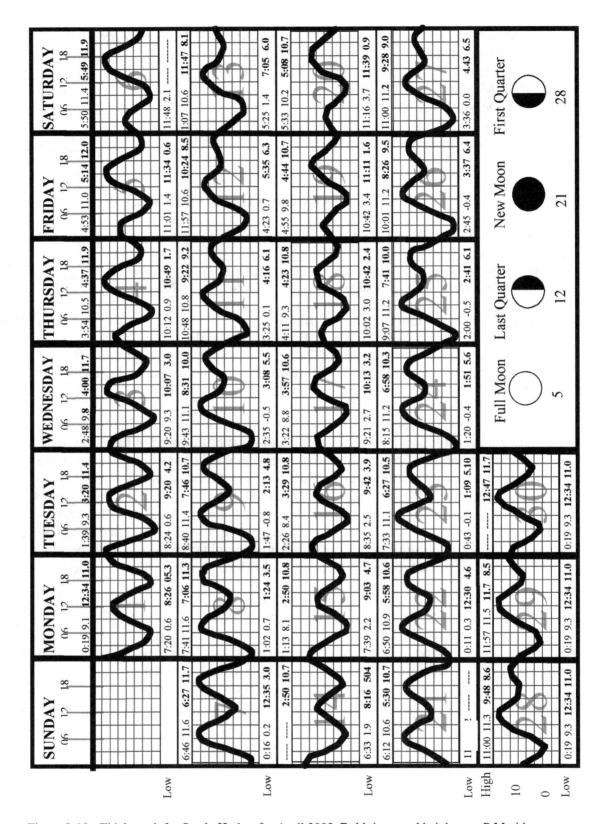

Figure 8-10. Tidal graph for Sandy Harbor for April 2002. Bold times and heights are P.M. tides.

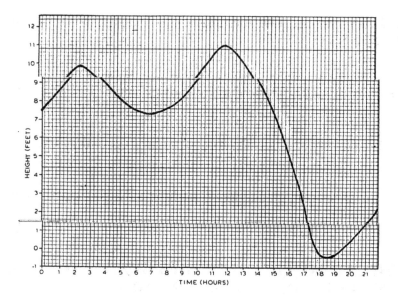

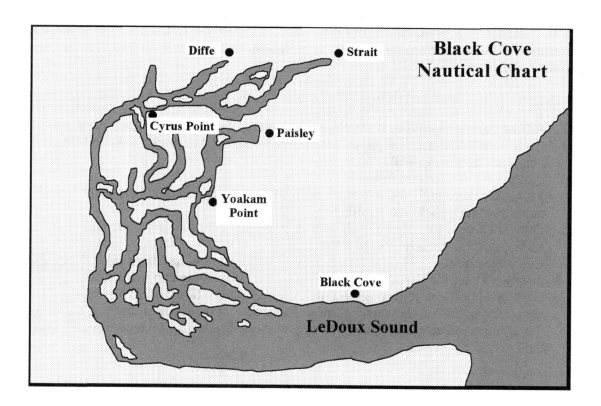

Figure 8-11. Black Cove map and tidal curve for April 24, 2007.

a.  How and why does the time and tide height change from Black Cove towards Diffe (Figure 8-11)?


b.  What is the difference in tidal range between Black Cove and Cyrus Point?


c.  Why is the tidal range so much larger at Strait than at Black Cove?


d.  Why are the times of high and low tides later at Strait than at Cyrus Point?


## 8-5.  GRAPHICAL METHOD FOR DETERMINING TIDE HEIGHT AT ANY TIME (ONE QUARTER—ONE TENTH RULE)

Some publications provide only times and heights of high and low tides. Using the graphical method described below and illustrated in Figure 8-12, it is possible to draw an approximate tidal curve from this data so as to determine tide height at any time during the tidal cycle.

Procedure

1.  Adjust data for location, if necessary, as done in Exercise 5 in this laboratory.
2.  On standard linear graph paper, place time along the horizontal axis and tide height along the vertical.
3.  Plot points of high and low tides.
4.  Join successive high and low tides with straight lines and then divide the lines into quarters. Mark each line's midpoint.
5.  Draw a vertical line above each quarter point adjacent to a high tide and below each quarter point adjacent to a low tide.
6.  Calculate one-tenth of tidal range between adjacent low and high tides. Mark this distance along the vertical lines at the quarter points between these tides.
7.  Draw a smooth curve through points of low and high tides, midpoints and measured points on the vertical lines. This curve approximates the tidal curve for the time period covered. Tide heights at various times may be read directly from the curve.

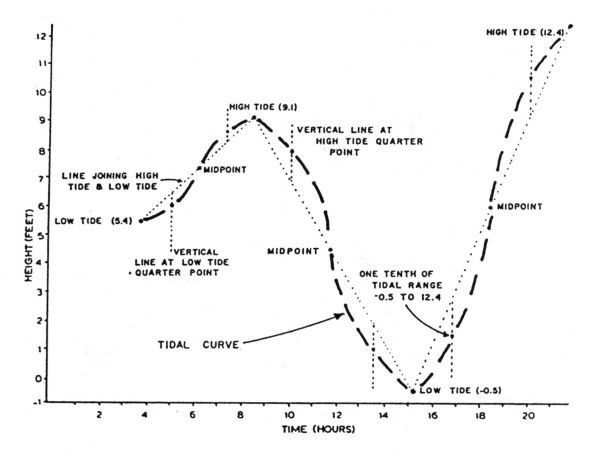

Figure 8-12. Demonstration of the graphical method for generating a tidal curve.

## EXERCISE 6.   CONSTRUCTING A TIDAL CURVE GRAPHICALLY

1. On the graph (Figure 8-13), draw a tidal curve for Paisley, using the One Quarter—One Tenth Rule and the data provided below.

   a. Data for Black Cove tides on April 19, 2007:

   | | | |
   |---|---|---|
   | *High Tides* | 5:49 AM | 11.3 ft |
   | | 4:12 PM | 10.9 ft |
   | *Low Tides* | 11:11 AM | 06.2 ft |
   | | 11:17 PM | −0.8 ft |

   b. Correction values for Paisley tides:

   | Time Correction | | Height Correction | |
   |---|---|---|---|
   | *High* | *Low* | *High* | *Low* |
   | +0 07 | +0 06 | +0.5 | 0.0 |

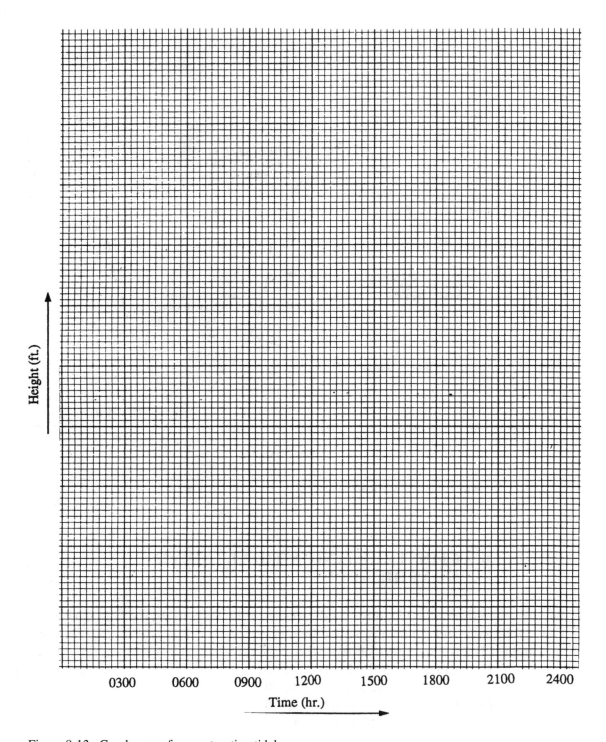

Figure 8-13.  Graph paper for constructing tidal curve.

2. After completing the tidal curve, determine the t ide height at each of the following times.

0900_____          1215_____          1500_____          2130_____
(9:00 AM)                (12:15 PM)              (3:00 PM)               (9:30 PM)

## General Questions

1. Is it possible to have three spring tides in 1 month? Why?

2. Do all months have the potential t o have three neap tides? If not, which ones  can not? Why?

3. Why would three neap tides or three spring tides in 1 month be rare?

4. Does the lowest low tide have to occur on the day of a spring tide? Does the highest high tide? Why?

5. Do tides only occur in the surface water or also below the pycnocline?

6. How does Coriolis effect differ from the centrifugal effect?

# Laboratory 9

# Nautical Charts 1 and Time

## 9-1. NAUTICAL CHARTS

**Nautical charts** are maps of the ocean or coastal areas. They provide information that is useful in navigation or piloting. Data commonly provided on charts include water depth, bottom characteristics, navigational aids, and dangers to ships. Some also provide information on currents and waves.

## 9-2. MAP PROJECTION

**Map projection** refers to transferring the surface of the spheroidal Earth to a flat sheet of paper. All charts have some distortion because it is not possible to transfer an image from a curved surface to a flat one without some areas being stretched and/or others compressed. Amount and location of distortion depends on the orientation and configuration of the sheet upon which the surface of the globe is transferred.

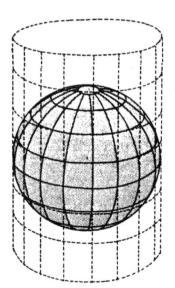

Figure 9-1. Construction of a cylindrical projection. Several lines of latitude and longitude have been projected onto the sheet.

Most nautical charts are **cylindrical projections**. In this method, a sheet of paper is rolled into a cylinder and placed tangent to the globe at some latitude. Points on the globe's surface then are projected onto the sheet in much the same manner as shadows would be cast if the globe was transparent and lit from the center (Figure 9-1).

On cylindrical projections, lines of latitude and longitude are straight and cross at 90°. Distortion of Earth's image rapidly increases away from the latitude at which the sheet is tangent to the globe. On a cylindrical projection of the entire Earth (called a **Mercator Projection**), the sheet is tangent along the equator and little distortion of Earth's image occurs in the equatorial regions. Distortion increases rapidly poleward and greatly exaggerates the size of polar areas. When measuring distances on a Mercator Projection, a different scale must be used at each latitude because of the rapid increase in distortion poleward. Despite this distortion, the great advantage to cylindrical projections is that straight lines plotted on the chart cut all meridians at the same angle and provide a compass course to follow that is also the shortest distance between two points and an arc of a great circle.

## 9-3.  DETERMINING LATITUDE AND LONGITUDE FROM CHARTS

There are a variety of tools to accurately determine the latitude and longitude of any point on a chart. The simplest and least expensive tool is used herein: a scale, also commonly called a ruler. Because nautical charts are based on cylindrical projections, not only are lines of latitude parallel, but all lines of longitude are also parallel. Thus, the same technique can be used on both.

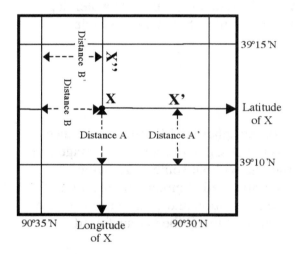

Figure 9-2. Determining latitude and longitude using a ruler.

From the point of interest (X) on the chart (Figure 9-2), draw a north-south line to the closest line of latitude. Measure the length of this line (distance A). Along the same line of latitude and on the same side, draw a north-south line of equal length (distance A′) to locate point X′. X and X′ are on the same latitude. Draw a line through X and X′ to the edge of the chart. Where the lines intersect the edge of the chart, read the latitude of point X.

From the point of interest (X) on the chart, draw an east-west line to the closest line of longitude. Measure the length of this line (distance B). Along the same line of longitude and on the same side, draw an east-west line of equal length (distance B′) to locate point X″. X and X″ are on the same longitude. Draw a line through X and X″ to the edge of the chart. Where the lines intersect the edge of the chart, read the longitude of point X.

## 9-4.  MILES-PER-DEGREE OF LATITUDE AND LONGITUDE

Most nautical charts do not contain a scale for measuring distance. Lengths on the chart are measured relative to latitude and converted into distance as statute (declared by law) miles, nautical miles, or kilometers.

Although lines of latitude are parallel, they are not evenly spaced because the Earth is a **spheroid**, not a perfect sphere. Earth's north-south axis is shorter than any east-west axis through the equator, indicating that the Earth bulges at the equator. The distance between each degree of latitude varies slightly, from 68.702 st. mi at the equator to 69.407 st. mi at the poles. Over most of the Earth, 1° change in latitude can be considered equal to 69.172 mi.

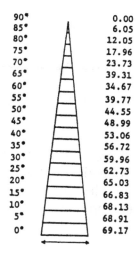

| | |
|---|---|
| 90° | 0.00 |
| 85° | 6.05 |
| 80° | 12.05 |
| 75° | 17.96 |
| 70° | 23.73 |
| 65° | 39.31 |
| 60° | 34.67 |
| 55° | 39.77 |
| 50° | 44.55 |
| 45° | 48.99 |
| 40° | 53.06 |
| 35° | 56.72 |
| 30° | 59.96 |
| 25° | 62.73 |
| 20° | 65.03 |
| 15° | 66.83 |
| 10° | 68.13 |
| 5° | 68.91 |
| 0° | 69.17 |

$$1° \text{ latitude} = 69.172 \text{ st. mi}$$

In contrast, the distance between each degree of longitude varies greatly with latitude because lines of longitude converge poleward (Figure 9-3). At the equator, the distance between each degree of longitude is 69.17 st. mi. At 45° north or south latitude, the distance is only 48.99 st. mi, and at the poles it is 0 mi. For most charts, distances should be measured by comparison to latitude, not longitude.

Figure 9-3. East-west decrease in distance between each degree of longitude between the equator and pole.

## 9-5.  DISTANCE AND SPEED

Distances on land are measured in statute miles or kilometers. A **statute mile** (**st. mi**) is equal to 5280 ft. On the sea, distances are measured in nautical miles. A **nautical mile** (**nm**) is defined as the distance between each minute of latitude. It is equal to 6080 ft, about 1.15 st. mi or 1.85 km. Because there are 60′ in each degree, the distance between each degree of latitude is 60 nm. When stating distances, it is necessary to indicate units (nautical miles, statute miles or kilometers).

$$1′ = 1 \text{ nm} \cong 1.15 \text{ st. mi} \cong 1.85 \text{ km}$$

The speed of an object in water is expressed in knots. A **knot** is equal to 1 nm/hr. To convert knots to kilometers per hour multiply by 1.85, and to convert to statute miles per hour multiply by 1.15. A common error is to report speed as knots per hour. Knots per hour is **acceleration**, the increase in speed with time. One knot per hour is 1 nm/hr/hr and this means that the object increases its speed by 1 knot each hour.

Some coastal nautical charts provide a logarithmic speed scale for determining the speed in knots necessary to travel a given distance in an allotted period of time. After determining the distance to be traveled and the time allowed for the trip, place a sheet

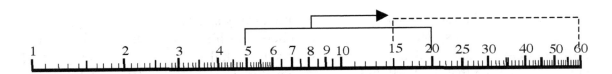

Figure 9-4. Logarithmic speed scale. To travel 5 nm in 20 min, a boat's speed must be 15 knots.

of paper along the speed scale and make marks on the paper corresponding to the number of nautical miles and time in minutes for the trip. Slide the paper so that one mark is on 60. The number corresponding to the other mark is the required speed in knots (Figure 9-4) to make the trip in the allotted time.

## EXERCISE 1.  LATITUDE, LONGITUDE, DISTANCE AND SPEED

1. Give the latitude and longitude of the following locations on the Cameron City chart (Figure 9-5).
   a.  Point David _____
   b.  Port Arbit _____
   c.  KTYU Radio Antenna _____
   d.  Ledge Rock _____

2. What is located near each of the following coordinates on the Cameron City chart?
   a.  31°13′00″N    131°48′30″W    _____
   b.  31°23′30″N    131°50′15″W    _____
   c.  31°14′00″N    131°52′45″W    _____
   d.  31°19′15″N    131°58′45″W    _____
   e.  31°18′30″N    131°53′30″W    _____
   f.  31°18′15″N    131°59′30″W    _____
   g.  31°22′00″N    131°52′30″W    _____
   h.  31°17′30″N    131°57′00″W    _____
   i.  31°09′30″N    131°49′50″W    _____

3. You are planning a boat trip from Cameron City to Ledge Rock, a distance of 5.5 nm. Using the logarithmic speed scale below, determine the speed you must maintain to make this trip in 20 min? _____
   What would your speed be if you had allotted 35 min for the trip? _____

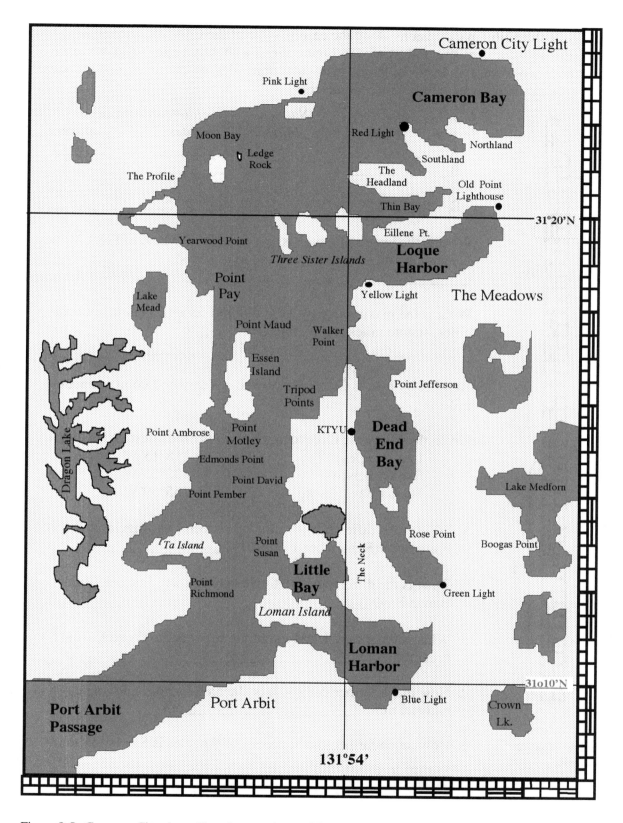

Figure 9-5.  Cameron City chart. Chart is not to be used for navigational purposes.

4. Note the latitudinal divisions along the eastern edge of the Cameron City chart (Figure 9-5). They are reillustrated in Figure 9-6 at a slightly larger scale. For each of the lettered areas, indicate its length in angular units (degrees, minutes, and seconds), nautical miles, statute miles and kilometers.

| Section | Angular Units | Nautical Miles | Statute Miles | Kilometers |
|---|---|---|---|---|
| A | _____ | _____ | _____ | _____ |
| B | _____ | _____ | _____ | _____ |
| C | _____ | _____ | _____ | _____ |
| D | _____ | _____ | _____ | _____ |

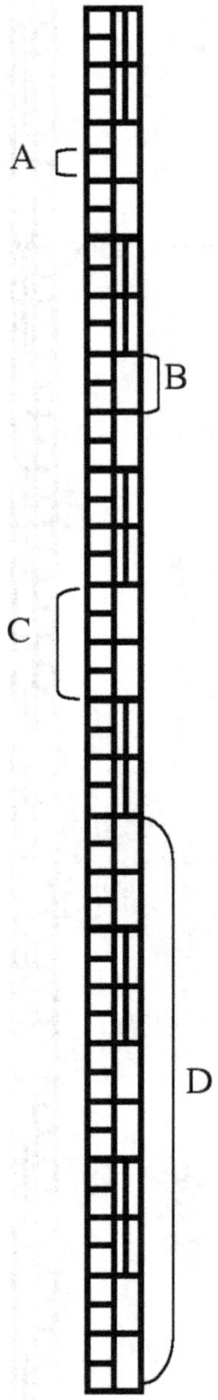

5. What is the length of Essen Island from Point Maud to Point Motley in nautical miles, statute miles and kilometers?
_____nm _____st. mi _____km

6. What is the maximum east-west width of Essen Island in nautical miles, statute miles and kilometers?
_____nm _____st. mi _____km

7. How far by boat is Cameron City from Yearwood Point in nautical miles? _____nm

8. Given the following three locations, answer the questions below.
Los Angeles             33°42′N  118°15′W
San Francisco           37°49′N  122°25′W
Hanna Bay, Hawaii       20°45′N  155°59′W

a. What is the difference in latitude between Los Angeles and San Francisco? (Remember 1° = 60′) _____

b. What is the difference in latitude between San Francisco and Hanna Bay? _____

c. What is the difference in longitude between Los Angeles and San Francisco? _____

d. How much farther due south is Hanna Bay from San Francisco in nautical miles? _____nm

e. How much farther due north of Los Angeles is San Francisco? _____km

A ⊏

B

C

D

Figure 9-6
Subdivisions of latitude.

## 9-6.  TIME

The solar day and year and the lunar month are the only natural observable units of time. The **solar year** is the amount of time required for the Earth to make one complete orbit around the Sun. Astronomically, it can be determined by observing the position of the stars. The **lunar month** is defined by the phases of the Moon and is equal to 28 days. The **solar day** is the amount of time required for Earth to make one complete turn on its axis. For early man, sunrise, noon and sunset were the only moments of the day that appeared to occur with regularity. Because sunrise and sunset varied with the seasons as the daylight lengthened and shortened, the solar day is measured from solar noon, the instant the Sun is at its zenith (highest point) in the sky. All other time units—millennium, century, decade, fortnight, week, hour, minute, and second—are artificial divisions with no scientific or observable basis.

Before mechanical clocks were reliable, most people depended on sundials to reset their clocks at noon each day. Each locality established its time on **local solar** time using a sundial. Because the Sun can be exactly above only one line of longitude (meridian) at noon, every other line of longitude lies either before noon or past noon. Clocks in nearby cities not on the same longitude showed different times because noon occurred slightly earlier or later at each location. As long as communication and transportation were slow, this posed no problem. With the development of telegraph and railroads, precise coordination of time became important. Initially, correction tables were published providing the minor time corrections between cities. Yet, as speed of communication and transportation increased, a better solution was needed. This led to the development of standard time.

By international agreement, **Standard Time** divides the Earth into 24 **time zones**, each encompassing 15° of longitude. All points within the time zone are considered to have the same time (Figure 9-7). Only the **central meridian** of a time zone has the correct solar time, i.e., the Sun is at its zenith above the central meridian when it is standard noon in that time zone. The difference between solar time and standard time increases away from the central meridian and reaches a maximum at either edge. Standard noon on the western edge of a time zone is 30 min early and on the eastern edge it is 30 min late. Central meridians are located every 15° east or west of the Prime Meridian. The boundaries between each time zone are 7.5° east and west of the central meridian. For political reasons and convenience, time zone boundaries are offset so as to avoid bisecting populated areas, some states and countries. Not all nations have adopted Standard Time and their official times differ from that of their time zone.

Located 180° from the Prime Meridian is the **International Date Line**, perhaps the most confusing part of Earth's time system. The International Date Line serves to separate that part of the Earth that has not yet finished 1 day from the part that has already begun the next day. When traveling across the date line, add one day if crossing from east to west

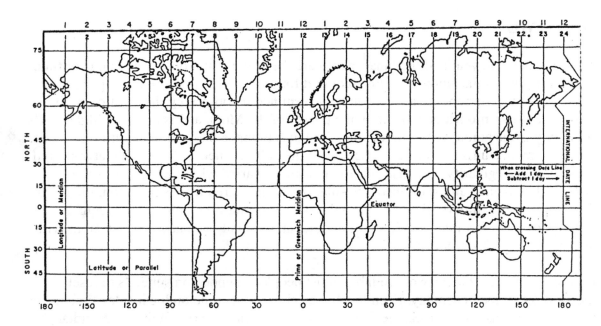

Figure 9-7. Mercator Projection of the Earth. Numbers at the top indicate time on a 12-hr clock. Those below give the equivalent time on a 24-hr clock. At the bottom are degrees longitude and on the left are degrees latitude. Longitudes shown are the central meridians of the time zones.

and subtract one day if crossing west to east. If it is Friday the 26th on the east side of the Date Line, it is Saturday the 27th on the west side.

The International Date Line is not the same as the location of midnight. Midnight and noon are fixed by the position of the Sun. Noon is at the line of longitude that is directly below the Sun and midnight is always 180° away, on the exact opposite side of the Earth from noon. As the Earth rotates and points pass across midnight, they begin a new day. The International Date Line is the line of longitude where we recognize the start of each new day. At any point on the Earth's surface, a day lasts for 24 hr, but each day lasts for 48 hr across the entire Earth. It takes 24 hr for every point on Earth to cross midnight and begin the day and an additional 24 hr for every point to reach midnight and begin the next day. Only when the Sun is directly above the Prime Meridian is every point on Earth experiencing the same day (Figure 9-8).

**(Standard) Daylight Savings Time** is a regional modification of standard time. During part of the year, time is advanced one hour or more so that more daylight hours are available in the evening. It is mainly used in the midlatitudes because of the seasonal differences in the number of daylight hours.

Although Earth's day is divided into 24 1-hr segments, most clock faces are numbered from 1 to 12. Hours from midnight to noon are labeled **AM** (*ante meridiem* = before noon) because these hours are after midnight and before the Sun is above the central meridian of the time zone and therefore before noon. Those from noon to midnight are

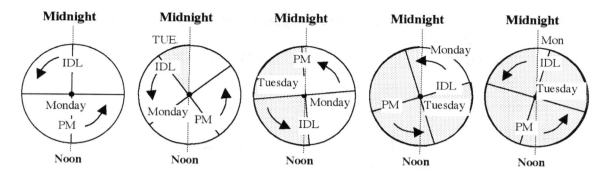

Figure 9-8. Progression of the day. When the Prime Meridian (PM) is at noon, the International Date Line (IDL) is at midnight and everywhere on Earth is the same day, Monday. As Earth rotates, the IDL crosses past midnight and Tuesday begins. Over the next 24 hr, the IDL rotates eastward until the PM is at noon and it is Tuesday across the entire Earth. It will require another 24 hr until the last part of Earth where it is Tuesday crosses midnight into Wednesday. Thus, a day lasts 48 hr across the entire Earth, but only 24 hr at any location.

labeled **PM** (*post meridiem* = after noon) because these hours are after the Sun was above the central meridian of the time zone. Noon and midnight are neither AM nor PM.

An alternative to the 12-hr clock is the **24-hr clock**, sometimes referred to as **military time** in the U.S, but more correctly known as **Universal Time**. Time on the 24-hr clock is expressed in terms of **hundreds**. Midnight is 0000 hr. 1 AM is 0100 (1 hundred hours) and 1 PM is 1300 (13 hundred hours). Minutes are recorded as the last two digits. For example, 7:58 AM would be 0758 and is read seven hundred and fifty-eight hours, and 7:58 PM is 1958, nineteen hundred and fifty-eight hours. Although the hours are stated as hundreds, each hour has only 60 min. To determine the time past noon, add 12 hr to change standard time to universal time or subtract 12 hr from universal time to change to standard time.

## EXERCISE 2.  TIME CONVERSION

1.  In the space below, list all of the central meridians (CM) in the western hemisphere and the boundaries (B) of their times zones.

<u>172.5E</u> <u>180</u> <u>   </u>   <u>   </u> <u>   </u> <u>   </u>   <u>   </u> <u>   </u> <u>   </u>   <u>   </u> <u>   </u> <u>   </u>
 (B)  (CM) (B)   (B) (CM) (B)   (B) (CM) (B)   (B) (CM) (B)

<u>   </u> <u>   </u> <u>   </u>   <u>   </u> <u>   </u> <u>   </u>   <u>   </u> <u>   </u> <u>   </u>   <u>   </u> <u>   </u> <u>   </u>
 (B)  (CM) (B)   (B) (CM) (B)   (B) (CM) (B)   (B) (CM) (B)

<u>   </u> <u>   </u> <u>   </u>   <u>   </u> <u>   </u> <u>   </u>   <u>   </u> <u>   </u> <u>   </u>   <u>   </u> <u>   </u> <u>   </u>
 (B)  (CM) (B)   (B) (CM) (B)   (B) (CM) (B)   (B) (CM) (B)

<u>   </u> <u>0</u> <u>7.5E</u>
 (B)  (CM) (B)

2. If it is 9:00 AM in New Orleans (90°W longitude), what time is it in each of the following locations?

   a. 135°E longitude _____     d. 9°W longitude _____

   b. 170°W longitude _____     e. Western New Zealand _____

   c. Greenwich, U.K. (0°) _____     f. Florida _____

3. If it is 10:25 PM in Cairo, Egypt (30°E), what are the ideal zonal borders for zones where it is the following times? Give answers in longitude.

| Time | Western Boundary | Eastern Boundary |
|------|------------------|------------------|
| 9:25 AM | _____ | _____ |
| 1:25 PM | _____ | _____ |
| 6:25 AM | _____ | _____ |
| 10:25 AM | _____ | _____ |
| 12:00 Noon | _____ | _____ |

4. Why does time change with longitude, but not with latitude?

5. If it is 7:00 AM on Monday, February 20, just east of the International Date Line, what are the time, day and date just west of the line? _____

6. Convert the following to time on a 24-hr clock.

   10:10 AM _____     1:15 PM _____     12:00 Noon _____

   12:04 AM _____     3:15 PM _____     12:00 Midnight _____

7. Convert the following to time on a 24-hr clock after adding or subtracting the minutes indicated.

   7:21 AM – 20 min _____

   1:05 PM – 37 min _____

   11:19 AM + 15 min _____

   8:05 PM + 77 min _____

8. Convert the following to time on a 12-hr clock.

   1345 _____     1723 _____     1157 _____

   0000 _____     0819 _____     2311 _____

   2121 _____     2400 _____

9. What is the difference between solar time and standard time?

10. Given the standard time at any location on Earth, is it possible to determine where it is standard noon?

    From the same information, is it possible to determine at what longitude it is solar noon?

11. Determine the time difference for each degree, minute and second of longitude.

    $1° =$ _____        $1' =$ _____        $1'' =$ _____

12. If it is 1930 solar time in San Francisco (122°25′W), where is it solar noon? Give the longitude. _____

13. If at any point on Earth a day lasts for 24 hr, how long does a day last across the entire Earth's surface? Explain your answer.

14. To make the excessively long movie "Titanic" not seem excessively long in advertisements its length was stated as 2 hr and 77 min.
    How long was the movie in hours and minutes? _____
    If the movie started at 1900, what time did it end? _____

# Laboratory 10

# Marine Ecosystem

## 10-1. COMPONENTS OF A MARINE ECOSYSTEM

An **ecosystem** is a group of species interacting with each other and with the environment in which they live. It involves the interaction of the biotic (living) and abiotic (nonliving) elements of the environment. **Abiotic** elements are all nonliving parts of the ecosystem, such as nutrients, heat, sunlight, the water, rocks, minerals, and dissolved salts and gases. The **biotic** component is all organic matter, including living organisms.

In most ecosystems, three categories of organisms are generally recognized: plants, animals, and decomposers. Plants are **autotrophs** (**primary producers**) and possess the ability to produce their own food using the abiotic components of the ecosystem. Animals and decomposers are **heterotrophs** (**consumers**). They consume and use the tissues of other organisms as food, but return most of what they eat to the abiotic part of the environment. This material can then be recycled by the plants into new organic matter.

Animals may be divided into three categories: herbivores, carnivores, and omnivores. **Herbivores** feed on plants, **carnivores** eat only animals, and **omnivores** consume both plants and animals.

**Decomposers** are bacteria. They primarily break down the organic remains of plants and animals into simple inorganic material and release it into the environment where it can be recycled by the plants.

## 10-2. ENERGY FLOW, TROPHIC LEVELS AND ECOLOGIC EFFICIENCY

For all but a few ecosystems, the Sun is the ultimate source of energy. Through **photosynthesis**, plants store a small portion of the solar energy reaching Earth in the organic compounds they manufacture. Some of this energy is passed to the herbivores and omnivores (**primary consumers**) when they consume the plants. The energy gained by the consumers may be used in various metabolic processes or stored in the organic compounds of their bodies. A portion of this stored energy, in turn, will be passed on to the carnivores or omnivores when the primary consumers are themselves consumed.

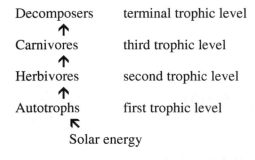

Figure 10-1. A simple food chain composed of several trophic levels.

Each level of production and consumption is called a **trophic level** (Figure 10-1). Plants form the first trophic level, herbivores the second, carnivores form the subsequent levels, and decomposers are the terminal level. Omnivores can be part of all but the first and last trophic levels because they eat both plants and animals. Much energy is lost between trophic levels because only a small portion of the energy stored in food is used to build new organic structures (growth) in the consumer. Most of the energy is expended in body maintenance, movement and production of indigestible structures (scales, bone, teeth, etc.), which provide no energy or nutrients when consumed at the next trophic level.

**Ecologic efficiency** is the percentage of energy that passes from one trophic level to the next and is used for growth. Because energy is difficult to measure, it is more common to discuss ecologic efficiency in terms of biomass. **Biomass** is the total mass (weight) of organic tissue present within a group of living organisms.

Ecologic Efficiency = Biomass of level (x + 1)/Biomass of level x

Ecologic efficiencies are highly variable and depend on age and nourishment of the organisms. Younger and poorly nourished organisms have higher efficiencies. This means that more of what they consume goes into growth. For natural marine ecosystems, 10% is generally accepted as an average efficiency. This mean that 90% of the energy consumed is expended by the organism and only 10% is used to generate new biomass that will be available for consumption at the next trophic level. At each trophic level, there is considerably less energy and biomass present than there was in the previous level.

## 10-3.  FOOD CHAINS, FOOD WEBS, AND BIOMASS PYRAMIDS

A **food chain** is a feeding sequence of organisms through which energy and matter is transferred from one trophic level to the next. Plants form the base of most food chains, herbivores, omnivores and carnivores the middle and decomposers the end. Because of low ecologic efficiencies, food chains are rarely longer than six or seven trophic levels.

The lower the trophic level at which an organism feeds, the greater the biomass available for consumption. For example, blue whales are the largest creatures on Earth today. They could not find sufficient food if they did not eat very low on the food chain. Most animals rarely feed on only one food source or one trophic level. This feeding diversity helps to ensure survival if one food source is in short supply. Also, animals unintentionally feed from several levels of the food chain because their food is frequently covered with bacteria and contains parasites. For example, herbivores accidentally consume large quantities of bacteria when they consume phytoplankton because the bacteria were living on the phytoplankton. By ingesting the phytoplankton, the herbivores are on the second trophic level, but by consuming the bacteria, they are also on the third or higher level. Omnivores can feed from several different trophic levels using a great many food sources. Carnivores normally feed from the third trophic level or higher.

Simple food chains quickly become complex and intertwined, forming a **food web**, a series of interconnected food chains. Most food webs are highly complex and it is frequently not possible to determine all of the chains within a web or how the chains are interlinked. All food chains in the ocean and on the land are interwoven to form a gigantic and complex global food web.

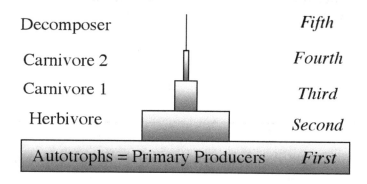

Figure 10-2. A biomass pyramid.

The **biomass pyramid** (Figure 10-2), also called the **energy pyramid** when energy and not biomass is considered, is a graphical representation of the amount of biomass at each trophic level of a food chain. As such, it also is a graphical representation of ecologic efficiency within a food chain. It must be stressed that biomass is not the same as the number of organisms. One kilogram of biomass could be one million phytoplankton or one small fish, but would be insufficient material to form the eye of an adult blue whale. Autotrophs always form the broadest portion of the pyramid in a stable food chain. If it is not sufficiently broad to support the upper part of the pyramid, the organisms in those higher trophic levels will begin to starve and die until their populations are reduced to a point where there is sufficient autotrophic biomass to support the survivors.

## EXERCISE 1.  A FOOD CHAIN

Using the data provided in this exercise, construct two simple food chains. Draw arrows between the members of each food chain to indicate the direction in which matter and energy flows. Identify the autotrophs, consumers, and decomposers of each chain. Label the trophic levels.

1.  Chain one.
    Diatoms (microscopic plants) are eaten by krill, which are consumed by a blue whale, which dies in a fjord and sinks to the bottom, where it is digested by bacteria.

2.  Chain two.
    Diatoms are eaten by copepods, which are consumed by herring, which become a meal for mackerel, which are ingested by swordfish, which becomes an expensive dinner for a man in a restaurant, who goes swimming and is eaten by a shark, which dies, washes ashore, and is decomposed by bacteria.

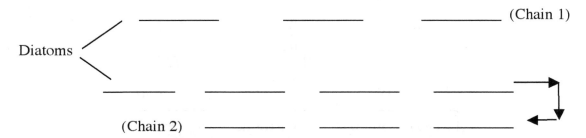

## EXERCISE 2.  A SIMPLE FOOD WEB

Using the data provided below, construct a food web. Draw arrows between the members of the web that indicate the direction in which matter and energy flow.

Diatoms and bacteria are consumed by molluscan larvae, copepods, tunicates, euphausids, cladocerans and bacteria. Cladocerans, euphausids, molluscan larvae, and bacteria are consumed by herring, amphipods, and bacteria. Copepods, molluscan larvae and bacteria are consumed by sand eels, arrow worms, and bacteria. Amphipods and bacteria are consumed by herring. Herring are consumed by bacteria.

Diatoms

# EXERCISE 3.  A BIOMASS PYRAMID

1. Using the data provided in Exercise 1, construct a biomass pyramid for the simple food chains 1 and 2. Label the trophic levels. Assume a 10% ecologic efficiency.

2. If 10,000 metric tons of phytoplankton is generated, what is the maximum possible biomass that will be present at each trophic level for the simple food chains above?

3. How much autotrophic biomass would be required to support 1000 tons of biomass at the 10th level of a food chain with 10% ecologic efficiency?

# EXERCISE 4.  CHANGING PYRAMIDS

Each year in the Antarctic region, there is exceptionally high productivity of phytoplankton (plants) in the spring and early summer because of the great increase in sunlight. This is reflected in the rapid increase in the population of herbivores and carnivores from spring through summer. In fall, phytoplankton production declines sharply as sunlight decreases. The number of herbivores and carnivores drops precipitously because of starvation and death. Only a relatively small population of autotrophs, herbivores, and carnivores are present in the winter.

1. In the space below, draw representative biomass pyramids showing autotrophs, herbivores and carnivores for the Antarctic region for the times indicated.

            Winter                        Spring/Early Summer

            Summer                    Fall

2. Using these pyramids, explain the changing abundance of carnivores and herbivores.

## General Questions

1. A blue whale weighing 10 metric tons feeds at the second trophic level. Assuming a 10% ecologic efficiency, how much biomass was required to produce this whale?

2. If this whale had fed on the third trophic level, how much first level biomass would have been required to produce the whale if ecologic efficiency was 10%?

3. Killer whales feed on whales and are on the fourth trophic level. How much first level biomass would be required to produce a 10 metric ton killer whale?

   How much blue whale could have been produced from this same mass if it fed at the second trophic level?

4. Why is it not surprising that the largest animals in the sea feed at the lower levels on the biomass pyramid?

5. Phytoplankton biomass provides a good measure of the biomass in the entire food chain or web. If the first level biomass in an area of upwelling is consistently 100,000 kg and the ecologic efficiency is 10%, what is the maximum biomass for all levels in a seven level food chain (to the nearest decimal point)?

# Laboratory 11

# Coastal Areas and the Shoreline

## 11-1. THE SHORELINE

The **shoreline** is the dynamic boundary where the land, sea and atmosphere meet. It is greatly influenced by winds, waves, tides, currents, activities of organisms, and long-term changes in sea level. These interact to modify the coast and develop distinctive coastal features.

## 11-2. LONGSHORE DRIFT

Waves approaching the coast refract shoreward as they enter water shallower than wave base (Figure 11-1). Although waves break nearly parallel to the beach, they strike the shore at an angle. The water from a breaking wave rushes diagonally up the beach face in the direction the waves are traveling until wave energy is spent. This water then flows straight down the beach face towards the sea. Some of this water is then caught by the next breaking wave and moved along the shore: diagonally up the beach face and then straight down. The zig-zag movement of sediment transported by this water along the beach face is called **longshore drift** and the

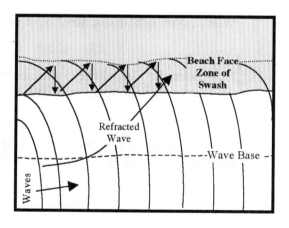

Figure 11-1. Longshore drift in zone of swash.

area across which the breaking wave washes is called the **zone of swash**. Because the location of the zone of swash changes with the tides, longshore drift transports sediments across a wider area of beach than is observed at any one time.

## EXERCISE 1.  LONGSHORE DRIFT

Procedure

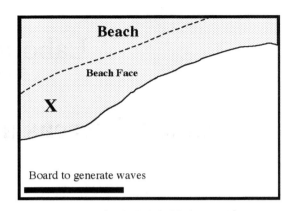

Figure 11-2.  Beach configuration for Exercise 1.

1.  In a large container (the larger the better), construct a beach, as shown in Figure 11-2, of light colored sand. The beach face should be a very gentle slope.
2.  Add some dark colored sand in the area labeled "X" on the figure.
3.  Use a board to generate waves on the side of the container opposite the beach where the beach extends "seaward." Waves should be large enough to wash across the beach face and move the dark sand.
4.  Continue generating waves until the dark sediment is distributed along the beach.
5.  Draw arrows indicating movement of dark sand along the beach and indicating direction of longshore drift.

Questions

1.  How did the waves transport the sediment?


2.  Was movement of sediment along the shore rapid? Why?


3.  How does the size of the container limit the movement of the sediment and development of longshore drift?


4.  If waves were generated from the right side of the container, would the sediment motion have been different? How? Why?


## 11-3.  COASTAL WAVE-INDUCED CURRENTS

**Longshore currents** are wave-generated currents that flow parallel to the shoreline (Figure 11-3). As waves break, they transport water to the shore. Before all of the water drains seaward, the next wave breaks and piles additional water against the land. Gradually, the sea level at the shore rises as more and more water is transported there. This is called **wave setup**. The higher the wave, the greater the shoreward transport of water and the higher the sea level at the shoreline. Along an irregular coast, the sea level will be higher where waves are larger and lower where they are smaller. The longshore current that develops is the downslope flow of water parallel to the shore from areas of larger waves towards areas of smaller waves. Where opposing longshore currents collide, they merge and turn seaward as a swift-flowing **rip current**. Because large waves are produced by refraction at headlands,

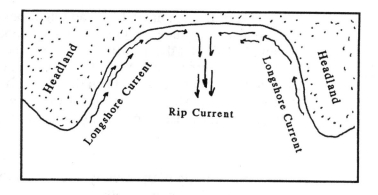

Figure 11-3. Longshore and rip currents between headlands.

longshore currents typically flow parallel to the shore away from the headland and towards intervening coves.

For long straight coasts, incoming waves push the accumulating water from wave setup along the coast until there is sufficient mass to form a rip current that can flow seaward against incoming waves. Longshore currents transport considerable sediment along the shoreline and rip currents carry sediments offshore.

## EXERCISE 2.  LONGSHORE CURRENTS AND RIP CURRENTS

Procedure

1.  In a large container (the larger the better), construct a shoreline as shown in Figure 11-4. Ideally the container should have a white bottom, or be clear and placed over white paper.
2.  Add sufficient water so that the waves will be deep-water waves where they originate.

3.  Using a short board, begin to generate waves by moving the board rapidly up and down near the end of the container opposite the beach. Continue to generate waves throughout the exercise.
4.  Locate the area offshore from the headland where the waves first begin to bend toward the shore. This is the location of wave base. Begin to add drops of blue food coloring in this area. As each drop disperses in the water, add another. Observe what happens to the coloring.

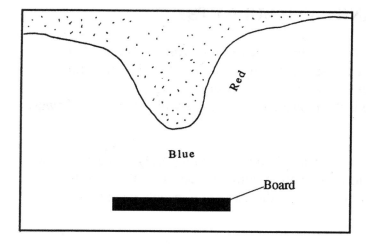

Figure 11-4.  Shoreline to be constructed in Exercise 1.

5. To one side of the headland, begin adding drops of red food coloring. As each drop disperses, add another. Observe what happens to the coloring.

6. On Figure 11-4, plot the movement of the two colors.

7. Continue generating waves until the coloring has moved along the coast and begins to be distributed away from the beach. If no motion occurs try generating larger waves by pushing the board forward slightly with each upward motion.

    a. What initially causes the coloring to be dispersed?

    b. Does the coloring move rapidly away from the beach or remain near the shore where it was introduced? Why?

    c. Locate where the waves are the largest and the smallest along the coast. Why are they located in those areas?

    d. Is there a relationship between the size of the waves and the direction the coloring flows? Explain.

    e. Did rip currents develop? How are they identified?

    f. How is this system unlike a real shoreline? How will this alter the results?

## 11-4. BEACH PROFILES AND SAND BUDGETS

Beaches can be divided into several parts. **Nearshore** extends from where breakers first begin to form to where waves swash across the beach (Figure 11-5). The **backshore** continues landward from the zone of swash to where the sea no longer strongly influences the land. The **beach face** is the intertidal zone across which the zone of swash migrates with the tides.

A **beach profile** is a cross section of the beach aligned perpendicularly to the shoreline. The profile plots change in beach surface elevation with distance from the shoreline. By comparing profiles made at different times along the same line, beach expansion or erosion can be documented.

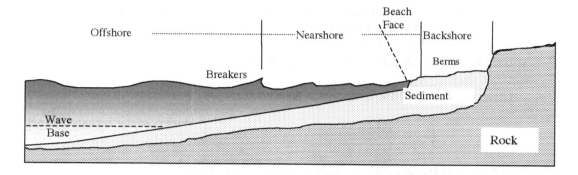

Figure 11-5.  Beach profile.

Beaches typically display seasonal variations in response to changing weather. Winter storms generate large waves, which are highly erosive. Sediment from the backshore is eroded and deposited in the nearshore area, producing a broad, flat beach face. Coarser sediments remain at the shore because they are too large and heavy for the waves to remove. After several storms, beaches tend to consist of coarse sediment and have a gently sloped beach face. This is a **storm profile**. In summer, storms are less frequent. The smaller waves move sediment shoreward. The coarse sediment gradually is buried by the finer sediment and the beach face becomes more steeply inclined. This is a **swell profile**.

A **sediment budget** is an estimate of the sediment added to and removed from a beach. Sediment sources include rivers, sea-cliff erosion, wind deposition, smaller waves, and longshore currents. In contrast, sediment is removed by longshore currents, large waves, and wind erosion. If more sediment is removed than added, the beach erodes. This is a **negative sediment budget**. If more sediment is added than removed, the beach expands. This is a **positive sediment budget**. If erosion and addition are balanced, the beach remains unchanged and is said to be in **steady state (dynamic equilibrium)**. It should be noted that although the beach does not appear to change when in a steady state condition, the sediment composing the beach is changing. As sediment is eroded from one area, it is replaced by sediment from another.

## EXERCISE 3.  BEACH PROFILES AND SAND BUDGETS

Procedure A

1.  In a rectangular container with clear sides, construct a beach with a cross section as illustrated in Figure 11-6. Use a mixture of fine sand to pebbles. This is the equivalent of a summer beach (swell profile).

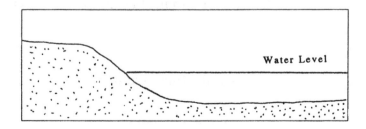

Figure 11-6.  Beach profile for Exercise 2.

2. Using an erasable marker, trace the beach profile you have constructed on the side of the container and then onto a sheet of paper. Label the various parts of the beach present in this model.
3. At the end of the container opposite the beach, generate small waves for several minutes.
   a. Describe how (if) the beach was altered.



   b. Where was the greatest impact of the waves?  Why?



   c. Which size sediment moved the most?  Why?



## Procedure B

1. Generate several large waves by pushing forward slightly as the board is raised.
   a. Describe how (if) the beach was altered.



   b. Where was the greatest impact of the waves?



## Procedure C

1. Continue to generate large waves until the beach ceases to undergo additional major change.
2. Using the marker, superimpose the new profile on the side of the container and copy it onto your original drawing.
3. On your diagram, indicate areas of erosion and the areas of deposition.
4. Describe how the new storm (winter) profile produced by large waves differs from the summer profile of the original beach.
   a. What type of sediment is most abundant at the shoreline?



   b. Where was the finer sediment transported?

## General Question

1. Prepare a sand budget for the beach by comparing the two profiles. List possible losses and additions.

2. Some beaches routinely shrink during the winter and expand during the summer. How is it possible to determine if they are in steady state?

## 11-5. BARRIER ISLANDS

**Barrier islands** are large, emergent deposits of sediments separated from the mainland by a shallow body of water called an estuary or lagoon. Between adjacent barrier islands there are narrow channels, called **tidal channels**, through which the tides flow. As sea level rises, barrier islands migrate shoreward. Storm waves wash over the low areas of the islands and transport sediment from the seaward side to the landward side of the island. This sediment is deposited as **washover fans** and forms the **back island flats**. Barrier islands can also grow vertically as wind-driven sands form sand dunes.

## EXERCISE 4.  BARRIER ISLANDS

Procedure A

1. Using the container from Exercise 2, construct a shore line and barrier beach with fine sands, as shown in Figure 11-7. Unnumbered contour lines are provided to indicate relative height of the land.
2. Generate small waves from the end of the container opposite the beach and observe what occurs as they reach the barrier islands.
   a. Compare the size of the waves that break at the barrier island and those in the lagoon.

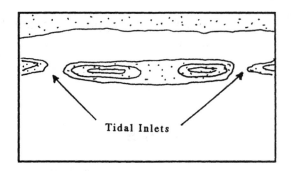

Figure 11-7. Shoreline design for Exercise 4.

   b. Why are waves in the lagoon smaller?

Procedure B

1.  Generate several medium size waves and observe what occurs as they reach the barrier island.
    a.  Are the waves in the lagoon still smaller than those reaching the seaward side of the barrier island? Why?

    b.  In what direction, if any, is the sediment of the barrier beach being transported?

Procedure C

1.  Generate several waves large enough to partially washover the barrier island in the area between the two "sand dune" hills.
    a.  In what direction was the sediment between the hills transported?

    b.  Did this change the profile of the back of the barrier island?

    c.  What name is applied to broad, flat deposits of sand on the backside of the island?

## 11-6.  ESTUARIES

An **estuary** is a semi-enclosed shallow body of water where fresh water from rivers and sea water mix. Estuarian, also called estuarine, circulation is strongly controlled by river outflow and tidal currents. In a **salt-wedge estuary**, river flow is more important than tidal mixing. Well-defined stratification develops between the low-density fresh water at the surface and the denser sea water below. As the river water flows seaward, internal waves are produced on the halocline. When these waves break, they incorporate small quantities of salt water into the fresh. Replacement of this salt water from below, together with the friction of the seaward-flowing surface water, generates a slow-moving, landward current in the salt water at the bottom.

In a **partially-mixed estuary**, water stratification is greatly weakened by increased tidal turbulence. Salinity does not change significantly within the water column. Bottom water easily mixes into surface water and a strong landward bottom current is established as the surface water flows seaward.

# EXERCISE 5.  ESTUARIAN CIRCULATION

*SALT-WEDGE ESTUARY MODEL*

Procedure A

1.  Incline a long, narrow, clear-sided container, as shown in Figure 11-8, and place it so that the lower end can overflow into a drain. Place a white background behind the container.
2.  Fill the container to overflowing with light blue, highly saline water.
3.  Place a hose at the raised end of the container and allow fresh water to flow gently into the container through the hose. The hose represents the river flowing into the estuary.
4.  Observe what occurs as the fresh water flows across the salt water.
    a.  Is a halocline present? Does it appear to be very strong?

    b.  Do the two water masses mix easily? Why?

    c.  Are internal waves present? If present, why do they form?

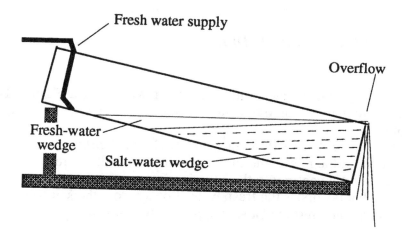

Figure 11-8.  Apparatus used to demonstrate salt-wedge and partially mixed estuaries.

Procedure B

1. Prepare a small quantity of dark red, highly saline water. Using a straw, inject this water at various heights in the saline water column below the fresh water.
   a. Describe what occurs.

   b. How quickly does the bottom water flow and in what direction?

   c. Do all of the red areas flow in the same direction? At the same speed? What does this indicate about water movement in the salt water wedge?

Procedure C

1. Increase the flow of fresh water so that larger internal waves are generated.
   a. Describe what occurs where the two water masses meet.

   b. How quickly does the bottom water flow now?

   c. Do all of the red areas flow in the same direction? What does this indicate about water movement in the salt water wedge?

## PARTIALLY-MIXED ESTUARY MODEL

Procedure A

1. Incline a long, narrow, clear-sided container, as shown in Figure 11-8, and place it so that the lower end can overflow into a drain. Place a white background behind the container.
2. Fill the container to overflowing with light blue, *slightly* saline water.
3. Place a hose at the raised end of the container and allow fresh water to flow through the hose gently into the container.
4. Observe what occurs as the fresh water flows across the salt water.
   a. Is a halocline present? Does it appear to be very strong?

b. Do the two water masses mix easily?

c. Are internal waves present?

Procedure B

1. Prepare a small quantity of dark red, slightly saline water. Using a straw, inject this water at various heights within the saline water column. Describe what occurs.
    a. How quickly does the bottom water flow and in what direction(s)?

    b. Do all of the red areas flow in the same direction? At the same speed? What does this indicate about water movement in the salt water wedge?

Procedure C

1. Increase the flow of fresh water so that larger internal waves are generated.
    a. Describe what occurs where the two water masses meet.

    b. How quickly does the bottom water flow?

    c. Do all of the red areas flow in the same direction? At the same speed? What does this indicate about water movement in the salt water wedge?

## 11-7.  DELTA CONSTRUCTION

A **delta** is an accumulation of river sediments deposited at the mouth of the river. Deltas form and grow seaward if the deposition of river sediments exceeds sediment removal by waves, currents and tides. As the delta grows, the river repeatedly changes its course and the location of sediment deposition. Initially, the river builds a long narrow delta (Figure 11-9), but during a flood, the river may erode a new channel with a steeper gradient in a different direction. The abandoned channel (called a **distributary**) gradually becomes clogged with fine sediment. Another long, narrow delta is constructed along the new channel, until it also is abandoned. Repeated shifting of the channel creates a broad mass of sediment crossed by numerous distributaries.

If large waves generate strong longshore drift, it will transport the river sediment along the coast and only a slight expansion of the shore may develop at the mouth of the river. Similarly, a large tidal range will produce strong ebb and flood currents and redistribute the sediments seaward, forming a series of sandbars and mudflats that radiate from the mouth of the river.

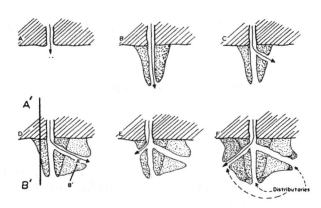

Figure 11-9. Growth of a delta. Cross section A′–B′ is shown in Figure 11-10.

The surface of the delta can be divided into three parts (Figure 11-10). The **prodelta** is the submerged, nearly horizontal, outermost part of the delta. It consists of nearly horizontal, clay- and silt-rich **bottom-set beds**, which grade seaward into the sediments of the continental shelf. The **delta front** is the broad, submerged, sloping face of the delta. It consists of the cross bedded (inclined layers) silts and sands of the **fore-set beds**. These beds usually comprise the bulk of the deltaic deposit. Sediments collect at the top of the delta front, become oversteepened and slide down the delta front and onto the top of the bottom-set beds. The **delta plain** is the flat, low-lying top of the delta. Part of it is emergent. It consists of thin, nearly horizontal **top-set beds** laid atop the fore-set beds. Additional layers are added to the top-set beds whenever the river overflows and floods the delta plain. As the delta builds seaward, fore-set beds bury the bottom-set beds, and are in turn covered by the top-set beds.

## EXERCISE 6.  DELTA CONSTRUCTION

Procedure A. Cross sectional view

1.  Incline a narrow, clear-sided container, as illustrated in Figure 11-11, so that it can overflow into a drain.
2.  Place a hose at the container's raised end. Allow water to flow through the hose and gently into the container.

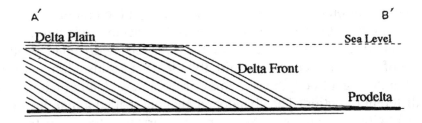

Figure 11-10.  Cross sectional view through a delta along the line A′–B′ on Figure 11-9.

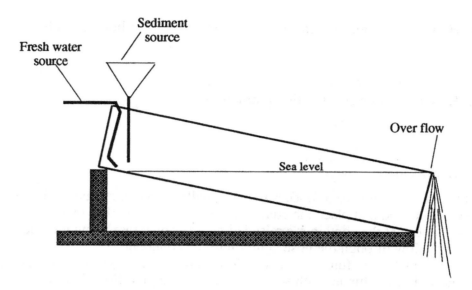

Figure 11-11.  Apparatus for the construction of a delta.

3. Install a small neck funnel so that its lower end is just in front of the mouth of the hose. Fill the funnel with sediment small enough to flow through the neck. The hose represents a river and the sediment is the river's sediment load.
4. Periodically add some moist powdered clay or moist chalk dust and some coarser granules directly to the water at the mouth of the hose.
5. Keep the funnel filled with sediment until a considerable mass has accumulated in the container and a delta has formed.
   a. Where is the sediment deposited as it flows into the standing water?

   b. Is all of the sediment deposited in the same location or is it being sorted by the current? Describe. Where is the coarsest sediment deposited? The finest?

   c. As the sedimentary mass develops, does the location of deposition change? Explain.

   d. Draw the cross section visible through the side of the container and label the following parts: top-set beds, fore-set beds, bottom-set beds, delta plain, delta front, and prodelta.

e. How are the fore-set beds deposited? Describe. The bottom-set beds?

f. How does delta construction promote sorting?

Procedure B. Plan view

1. Obtain a wide (at least a foot), relatively shallow container. Incline it slightly toward one end and place the container so that it can overflow into a drain.
2. Place a hose at the container's raised end, midway across its width, and allow water to flow through the hose gently into the container.
3. Install a small neck funnel so that its lower end is directly in front of the mouth of the hose. Fill the funnel with sediment small enough to flow though the funnel. The hose represents a river and the sediment from the funnel is the river's sedimentary load.
4. Keep the funnel filled with sediment until a considerable mass has accumulated in the container and a delta has formed. Periodically increase the flow of water moderately to simulate a flood and then return to the previous lower flow.
   a. Where is the first sediment deposited as it flows into the standing water?

   b. Is all of the sediment deposited in the same location or is it being sorted by the current? Describe. Where is the coarsest sediment deposited? The finest?

   c. As the sedimentary mass develops, does the location of deposition change? Explain.

   d. How does the river change its course as it flows across the delta plain?

   e. How does the delta change during "floods"? Describe.

   f. How does the delta plain become emergent?

# 11-8.  HUMAN MODIFICATION OF THE SHORELINE

People modify the shoreline to prevent erosion and improve areas for shipping, industry, recreation, and urban development. They build structures, dredges channels, and add or remove sediments from coastal areas. In addition to the initial, desired effects, unexpected and undesirable consequences frequently develop from human intervention.

Structures built to modify the coast can be divided into two groups: those parallel and those at an angle to the shoreline. Structures parallel to the coast include revetments, sea walls, bulkheads, and breakwaters. These mainly function to prevent erosion as waves strike the land or to have the waves expend their energy before reaching shore. Revetments, seawalls, and bulkheads are built directly against the land they protect. A **revetment** is a low sloping apron of concrete or rock fragments (**rip rap**), too large for waves to remove. They are placed at the foot of an easily eroded slope or at the base of another shoreline structure. **Seawalls** are structures designed to protect areas from high waves and prevent flooding and erosion. Seawalls and revetments commonly are used together because seawalls alone may be undermined by wave erosion. A **bulkhead** is a wall built to prevent land or landfill from sliding into the sea. Frequently, bulkheads are constructed offshore and the area they enclose is then filled with sediment.

Although revetments, sea walls, and bulkheads may temporarily protect the area behind them, they do not prevent and may even promote erosion of the beach or shallow sea floor in front of them. These artificial structures also may deprive adjacent beaches of their major source of sediment. This will produce a negative sediment budget and beach erosion. With time, these structures may be undermined and fail. Vertical seawalls at the water's edge can cause increased erosion elsewhere. Because vertical walls reflect waves, opposing shores will experience increased wave energy as both reflected waves and direct waves break on the shore. This increased wave activity can produce stronger longshore currents, which can erode more sediment from the beach.

A **breakwater** is a structure built offshore from the area it protects. Waves expend their energy against the breakwater and only greatly weakened, refracted waves reach the shore behind the structure. The reduced wave action protects the shore, but encourages sediment deposition behind the structure. With time, the beach expands and may eventually reach the breakwater. Areas downdrift from the breakwater may erode because they are deprived of the sediment trapped behind the breakwater.

Groins and jetties are constructed at an angle to the beach. **Groins** are low walls built out from the shore into relatively shallow water. They interrupt longshore drift and trap sediment on their updrift side. Groins are used to restore and expand beaches. Because groins trap sediment on the updrift side, there is no sediment to replace what longshore drift removes on the downdrift side. This side begins to erode.

Longshore drift can shift the location of tidal channels and the mouths of slow-flowing rivers by depositing sediment at the updrift edge and eroding it from the downdrift edge. The channel gradually migrates in the direction of drift and may become so filled with sediment that it will be too shallow for boats to pass. **Jetties** are structures built on the updrift side of a channel or river mouth so as to trap sediment, thereby maintaining the channel's depth and stabilizing the channel's location. Jetties are similar to groins, but extend farther offshore into deeper water. Sediment trapped on the updrift side of the jetty is not available to downdrift beaches. As the updrift beach expands, downdrift beaches erode. Pairs of jetties, one on each side of the channel, are sometimes used to confine the current through the channel, causing it to flow faster, thereby preventing sediments from being deposited within the channel. Jetties can be combined with a breakwater to form an artificial harbor.

The correct placement of groins and jetties is extremely important. Built too close to another structure, they will cause the structure to be undermined as the beach erodes on the downdrift side or to be stranded in shallow water on the updrift side. Placement is further complicated if wave direction, and resulting longshore drift, changes with the seasons. One side of the jetty or groin would be downdrift for part of the year, but updrift for the remainder.

**Dredging** to deepen channels and harbors produces depressions, which act as sediment traps and can disrupt the longshore transport of beach material. As sediment fills the dredged areas, downdrift beaches will be deprived of sediment and begin to erode. To maintain channel depth, dredging will become a periodic necessity. Dredging sediment from far offshore to replenish beaches should have little impact on the coast, but nearshore dredging can produce erosion.

Addition of sediment to a beach (**beach replenishment**) will temporarily expand the beach where the sediment is added, but longshore drift will immediately begin to redistribute the sediment and expand the beaches downdrift. Wharves and docks located downdrift may become stranded in shallow water. The additional sediment may become trapped in channels, bays, or behind jetties. Dredging may be required to remove this sediment. The life span of jetties will be shortened as the excess sediment rapidly fills the updrift side and sediment begins to be deposited around the end of the jetty. Eventually, there will be little evidence that the beach had been replenished where the sediments were initially deposited.

Actions taken far inland can indirectly modify coastal processes. When land is deforested and erodes, or when mining and farming increases the sediment load of rivers, additional sediment will gradually reach the coast and cause beaches to expand. In contrast, dams on rivers will trap sediment, decreasing the amount of sediment reaching the coast. Beaches will begin to erode as longshore drift and currents carry sediments away.

Summarized below are typical human modifications, their intended results and some possible negative consequences.

| Modification | Desired Result | Possible Negative Consequences |
|---|---|---|
| 1. Revetment, seawall, bulkhead | Protect land from erosion | Loss of beach in front of structure, possible reflection of waves to other areas, deprives downdrift areas of sediment |
| 2. Breakwater | Protects shore line, creates safe harbor for ships during storms | Traps sediment in quiet water, causes erosion downdrift |
| 3. Groin | Traps sediment to enlarge beach on updrift side | Erosion of beach downdrift, stranding structures updrift |
| 4. Jetty | Maintains location and depth of channels | Erosion of beach downdrift, stranding structures updrift |
| 5. Dredging | Replenishment of beaches, deepens channels and harbors | Beach expansion downdrift, stranding piers, deeper channels act as trap for sediment, erosion downdrift |
| 6. Dams on rivers | Water supply, prevent flooding, hydroelectric power | Less sediment from river, beach erosion |
| 7. Deforestation, mining | Different land use, recover resources | Excess sediment from river, beach expansion, stranding of structures, filling of channels and bays |

# EXERCISE 7.  HUMAN MODIFICATION OF SHORELINE

1. Hypothetical coast A (Figure 11-12)
   Waves shown are coming from the dominant wave direction.
   a. Draw arrows indicating the most likely direction of sediment transport for the various areas of the coastline shown.
   b. Identify the artificial structures along the coast and indicate their main objective.

   | Structure | Objective |
   |---|---|
   | i. _____ | _____ |
   | ii. _____ | _____ |
   | iii. _____ | _____ |
   | iv. _____ | _____ |

   c. The chart (Figure 11-12) shows the area immediately after the structures were built and before the shoreline adjusted to them. Indicate where sediment would be expected to collect and erode.

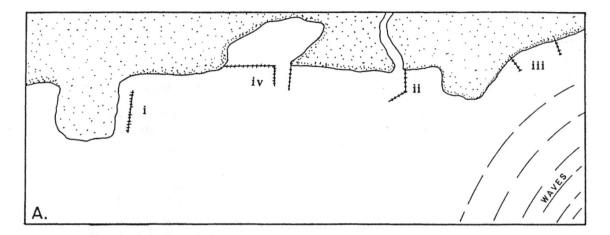

Figure 11-12. Hypothetical coast A.

2. Hypothetical coast B (Figure 11-13)
   Waves shown are coming from the dominant wave direction.
   a. Draw arrows on the chart to indicate the most likely direction of sediment transport for the various areas of shore illustrated.
   b. Identify the artificial structures along the coast and indicate their main objectives.

| | *Structure* | *Objective* |
|---|---|---|
| i. | _____ | _____ |
| ii. | _____ | _____ |
| iii. | _____ | _____ |
| iv. | _____ | _____ |

   c. The chart shows the area immediately after the structures were built and before the shoreline adjusted to them. Indicate where sediment would be expected to collect and erode.

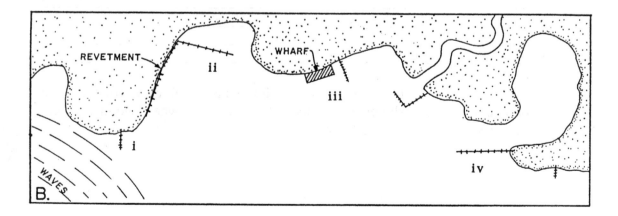

Figure 11-13. Hypothetical coast B.

3. Hypothetical coast C (Figure 11-14)
   As a coastal engineer, you have been hired by the state to develop a plan to protect the village, factory, docks and public beaches. Your study has revealed that the sea cliff on which the village is located erodes easily and is the major source of sediment for the beaches. Only a minor amount of sediment is supplied by the river. Major wave directions are indicated on the chart. Suggest possible structures to control the coast and indicate possible negative effects that could result from each.

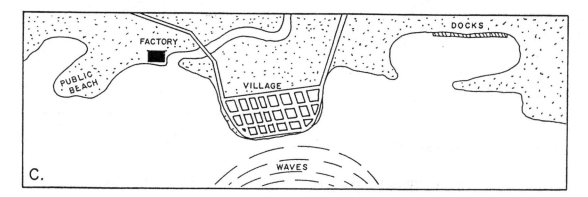

Figure 11-14. Hypothetical coast C.

4. Hypothetical coast D (Figure 11-15)
   As a coastal engineer, you have been hired to prepare a plan for the area shown on the coastal chart. The main goals are to protect the town on Mitchell Point and the harbor facilities in Herman Bay. The land area consists of easily eroded sediment, mainly sand and silt. A study of wave and wind records for the past ten years indicates that from October through May the wind and waves were primarily from the southwest. From June through September, they were from the southeast.
   a. Why are the wind and wave directions important?

   b. In red, draw arrows along the coast indicating the direction in which longshore currents and longshore drift would be directed during the summer. In blue, indicate the same for winter. Remember to allow for refraction around headlands. Do rip currents occupy the same position year round or do they change location with the season?

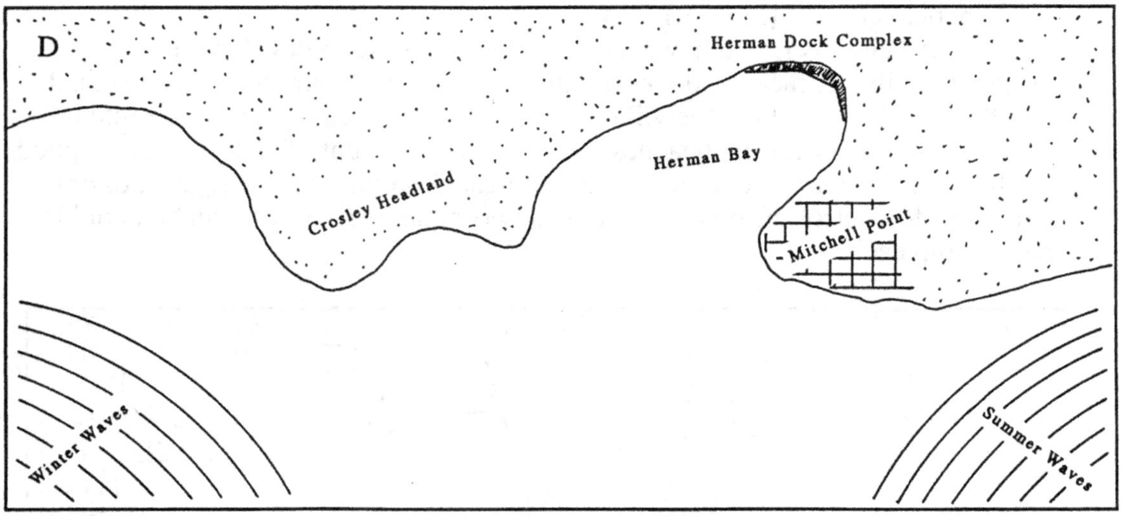

Figure 11-15. Hypothetical coast D.

    c. Identify those areas, if any, where the currents do not change over the course of the year.

    d. In some areas, longshore drift is to the right for six months and then drift is to the left for six months. Yet, over the year the sediment is transported more in one direction than in the other. Explain why this would occur.

    e. Why would Herman Bay be expected to eventually fill with sediment?

    f. Indicate what structures or activities you would propose to protect this coastal area from additional erosion. What possible negative consequences could result from your plan? Be specific. Do not just list general negative consequences from the summary earlier in this chapter.

5. Hypothetical coast E (Figure 11-16)

   As a town counselor, you must vote on a proposal to allow a major hotel and seaside resort to expand its beachfront by artificial replenishment (renourishment). The area of expansion is shown on the chart below. The hotel proposes to dredge sand from deep water far offshore. Because the sand is clean, there should be no pollution. The hotel's lawyers are arguing that the expansion will have little impact on the shoreline. Study the map and determine the possible long-term impact of the proposal. How would you vote on this measure? Justify your vote.

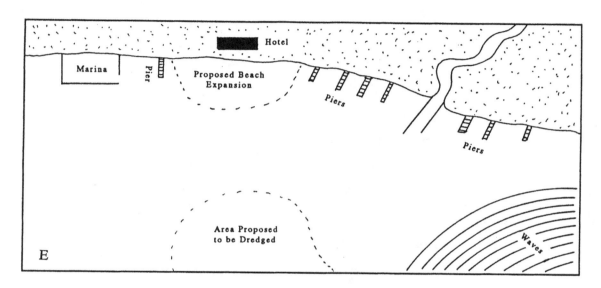

Figure 11-16.  Hypothetical coast E.

# Laboratory 12

# Ocean Pollution

## 12-1. FORMS OF POLLUTION

**Pollution** occurs whenever the activities of man cause concentrations of substances in the ocean to be raised above or depressed below "natural" levels, such that environmental conditions are markedly altered and can be considered deleterious to the living or nonliving parts of the ecosystem. The most common forms of pollution result from the release of hydrocarbons, sewage, heavy metals, radioactive elements, and biocides into the environment. Because these substances are too hazardous to use in lab, we will substitute safe materials that will simulate the behavior of these pollutants.

## 12-2. POLLUTION IN THE ENVIRONMENT

Most pollution enters the ocean from a **point source** and is then spread outward by currents, waves, and dispersion. Currents and waves can rapidly mix and spread the pollutant over a large area. This dilutes the pollutant and can greatly lessen its effects. Even in environments protected from waves and currents, pollutants will gradually be spread through dispersion, the random motion of molecules away from areas where they are concentrated.

## EXERCISE 1. DISPERSION OF MARBLES

Procedure

1. Place a cluster of 25 colored marbles in one corner of a tray and surround the cluster with 75 white marbles. The colored marbles represent pollution molecules and the white marbles represent water.
2. Begin to shake the tray randomly in various directions, allowing the marbles to roll about freely. Observe what happens to the colored marbles. Continue shaking the tray for about 2 min.

a. Did the colored marbles remain in a cluster? Why?

b. Does the distribution of colored marbles now appear to be random? Why?

c. Initially, the cluster of marbles in the corner of the tray was 100% colored, and colored marbles represented 25% of all marbles on the tray. Tilt the tray so that marbles again collect in the same corner. Take 25 marbles from the corner.
   What percentage is colored? _____
d. Tilt the tray to the opposite corner and again randomly collect 25 marbles.
   What percentage of these marbles is colored? _____
e. How effective was dispersion in diluting the concentration of the colored marbles?

## EXERCISE 2. DISPERSION IN WATER

Procedure

1. Following the instructions in Laboratory 5, prepare three tanks for water density experiments.
2. In the first tank, place dark blue, room temperature, fresh water on one side and clear, cold, fresh water on the other.
3. Carefully remove the barrier so as to minimize mixing and allow the water to settle. Through the side of the container, measure the thickness of the layers that form.

   Dark Blue_____        Light Blue_____        Clear_____

   What does the light blue layer represent? _____
4. Allow the container to sit undisturbed for about 40 min and again measure the thickness of the three layers. While waiting, proceed to step 5.

   Dark Blue_____        Light Blue_____        Clear_____

5. In the second container, place clear, room temperature, fresh water on one side and dark red, room temperature, salt water on the other.
6. Carefully remove the barrier so as to minimize mixing and allow the water to settle. Through the side of the container, measure the thickness of the layers that form.

   Dark Red _____        Light Red_____        Clear _____

   What does the light red layer represent? _____
7. Allow the container to sit undisturbed for about 40 min and again measure the thickness of the three layers. While waiting, proceed to step 8.

   Dark Red _____        Light Red_____        Clear _____

8.  In the third container, place light blue, room temperature, fresh water on one side and yellow, room temperature, fresh water on the other.

9.  Carefully remove the barrier so as to minimize mixing and allow the water to settle. Through the side of the container, measure the thickness of the layers that form. If the color banding remains vertical, measure their widths.

Light Blue_____        Green_____        Yellow_____

What does the green layer represent? _____

10. Allow the container to sit undisturbed for 40 min and then measure the thicknesses (or widths) of the three layers. (While waiting, proceed to the next part of the lab.)

Light Blue_____        Green_____        Yellow_____

a.  Assume that the colors red, blue, and green represents pollution from heavy metal, biocides or radioactive substances.

1.  As the barrier was removed, what caused the initial mixing?

2.  How rapid was this mixing?

3.  What caused subsequent mixing of the pollution (colors) into the clear water?

4.  How rapid was this subsequent mixing?

b.  Compare the results from the first two containers. In the first container, the mixing was between two fresh water bodies at different temperatures, but in the second it was between fresh water and salt water.

1.  Did they mix at the same rate?

2.  If at different rates, which mixed the more rapidly? Why?

3.  What is the implication for pollution in the sea?

4.  Which of the three containers displayed the greatest rate of mixing after 40 min? Why?

## 12-3.  HYDROCARBONS IN THE SEA

Most hydrocarbons are not readily soluble in water nor do they easily disperse into water. Because of their lower density, they generally float on the water in a variety of forms, from thin films to thick globules, depending on the nature of the oil. Some heavy oils are denser than water and sink to the bottom.

When oil spills occur, people attempt to control them with booms, dispersants and burning. **Booms** are floating barriers used to corral the oil and inhibit it from spreading so that it can be more easily recovered. **Dispersants** are chemicals that allow the oil to disperse as microscopic droplets into the water. **Burning** mainly removes the lighter oil, but creates air pollution, and leaves the heavier oils in the water.

## EXERCISE 3.  OIL SPILL RECOVERY AND DISPERSAL

Procedure

1. Partially fill a wide container with water and pour in enough olive or cooking oil to cover about 10% of the surface. This is the oil spill to be recovered.
   a. Describe what happens to the oil as it was poured onto the water.

   b. Is this a light, medium or heavy oil? _____
   c. Does it form a continuous layer across the water or several isolated pools?

   d. Stir the water. Blow some of the oil droplets together. How easily do separate pools merge?

2. One method of recovering oil from a spill is with a suction pump. A hose is lowered from a ship and used to suction as much oil (and as little water) as possible from the surface. Using an eye dropper or similar device as a suction pump, attempt to recover as much oil as possible from the spill in 1 min. You should count only the time while the dropper is sucking fluid. Place the recovered fluids into a small diameter glass container or small graduate cylinder and measure either the thickness of oil and water or their volumes. Record these data below. Pour the water and oil back into the spill and thoroughly clean and dry the glass container.

3. Another method of recovering oil is to use a skimmer, a devise that skims the surface of the water removing the layer of oil. We can duplicate this using a large spoon. Attempt to recover as much oil and as little water as possible from the spill in 1 min by depressing the bowl of the spoon into the oil and allowing the oil to flow into it. You should count only the time that the spoon is collecting fluid. Place the recovered fluids into the same glass container and measure the thickness or volumes of oil and water. Record these data below. Pour the water and oil back into the spill and thoroughly clean and dry the glass container.

4. Rarely is the sea surface calm. Waves of various sizes constantly churn the surface and make oil recovery much more difficult. With someone generating waves in the container, redo steps 2 and 3 and record the data in the spaces below. Waves should be of similar size throughout this experiment to validly evaluate the various techniques.

5. Booms are frequently used to concentrate the oil. Ships tow the booms around the spill and slowly begin to draw the ends of the boom together and concentrate the oil into a thicker pool. This can increase the efficiency of recovery. For this experiment, booms can be made by cutting soda straws into convenient lengths and sealing the ends with melted wax. A series of straw segments can be hinged together with short lengths of thread embedded in the wax. Using the booms, encircle the spill and concentrate it as much as possible. Repeat steps 2 and 3, first with no waves and then with waves. Again, uniformity of waves is necessary. Record the results below.

6. Generate a few large waves and note how effective booms are at containing the oil spill.

*Oil Spill Recovery Data*

Waves generated during recovery were:  small_____   medium_____   large_____

|  |  | No waves, no boom | Waves, no boom | No waves, with boom | Waves, with boom |
|---|---|---|---|---|---|
| Suction | Oil | _____ | _____ | _____ | _____ |
|  | Water | _____ | _____ | _____ | _____ |
| Skimmer | Oil | _____ | _____ | _____ | _____ |
|  | Water | _____ | _____ | _____ | _____ |

7. Evaluate the effectiveness of the two methods under the various conditions.
   a. How did waves alter the results?


   b. How effective were the booms when no waves were present?


   c. How effective are the booms when waves are present?

## EXERCISE 4.  DISPERSANTS

Dispersants do not remove oil from the water, but destroy the surface tension barrier that prevents the oil and the water from easily mixing. After using a dispersant, the oil is still in the water, but it is not visible. One of the most common dispersants is soap. That is why soap is useful in cleaning greasy dishes, washing clothes, and bathing. A given volume of dispersant can only disperse a limited volume of oil, depending on the type of oil and volume of water.

Procedure

1. Into each of four identical jars place about a half spoonful of oil and then fill the jars halfway with water. Label the jars A, B, C and D.
    To determine the effectiveness of soap as a dispersant, we must demonstrate that it is the soap causing the dispersion and not time or turbulence. Jar A will be the control jar. Once the oil is added, place a top on the jar and allow it to sit undisturbed throughout the experiment. Jar B will indicate the importance of turbulence alone in dispersing oil. It will be periodically shaken. Jar C will reflect the efficiency of soap alone in dispersing oil. Soap will be added to the jar, but it will not be shaken. Jar D will show the effect of soap and turbulence together.

2. Add about 10 drops of liquid soap detergent to jars C and D. Put tops on jars B and D and shake them vigorously for about a minute. Observe to see if oil is still present in all of the jars. If it is, add the same number of drops of soap to both jars C and D. Close jars B and D and shake vigorously again for about a minute. Repeat until oil vanishes from one of the jars. Compare the amount of oil in the other jars.
    a. From which jar does the oil vanish first? Why?



    b. Both jars C and D contain an equal amount of dispersant, but jar D was agitated. How important was agitation in dispersing the oil? Why?



    c. Would one expect the oil in jar C to eventually disperse without agitation ? Why?



    d. What effect did agitation alone have on the oil in jar B? Why?

e. Which of these four examples explains why an oil spill can travel great distances in wind-blown seas without much dispersion?

f. Would you expect the oil in jar A to ever disperse on its own?  Why?

## 12-4.  SEWAGE EFFLUENTS

The term **"effluent"** is neutral in its environmental meaning and simply refers to a fluid that is being introduced into the environment. Distilled water that is released into a stream would be an effluent, but not a form of pollution. Similarly, the term **"outfall"** only refers to where the effluent is allowed to enter the environment. Because we tend to hear the terms "effluent" and "outfall" used in describing pollution, the tendency is to think that all effluents are bad and that all outfalls should be eliminated. This is not the case.

**Sewage effluents** consist of complex mixtures of solids, semi-solids and dissolved material mixed in water. Frequently, the outfall is an underwater pipe terminating in a coastal sea at some distance offshore. As the effluent is released, the various components in the effluent will rise or sink depending on density. Because fresh water is less dense than seawater, the effluent usually rises toward the surface as a widening plume (Figure 12-1). This greatly expands the area affected by the pollution, but also dilutes the pollutant. As the plume spreads and mixes with the seawater, some material begins to sink. If a pycnocline is present, material may become trapped on that layer.

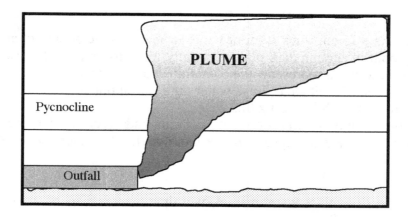

Figure 12-1.  Sewage effluent released below pycnocline. Intensity of shading indicates concentration of pollution.

Procedure

1. Following the instructions in Laboratory 5, arrange a container for a water density experiment. Place clear, room temperature, (very) saline water on one side of the tank and clear, room temperature, fresh water on the other. Remove the partition and allow the water to settle.
2. Using a soda straw or flexible plastic tubing, inject warm, turbid (muddy) fresh water at the bottom of the container. The straw represents the outfall and the turbid water the sewage. Observe what occurs through the side of the container.
   a. Where does the turbid water flow? Why?


   b. Over time, does material collect on the pycnocline? Why?


   c. After 5–10 min, generate waves on the surface and the pycnocline. What effect does this have on the pollution?


## 12-5. DREDGING

Dredging produces large quantities of sediment that must be discarded if there is no use for it. Dredging can also release volumes of contaminated **pore water**, the water trapped between the grains of sediment. Because of its long residence time, pore water tends to become saturated with any pollutants contained within the sediment. Release of the pore water can become a major source of pollution.


## EXERCISE 5.  DREDGING AND POLLUTION

Procedure

1. Place about a 15 cm-thick layer of white sand across the bottom of an aquarium.
2. Slowly pour in enough dark blue, very salty water to saturate the lower 14 cm of sand. This solution represents the contaminated pore water.
3. Slowly add sufficient clear, fresh water until it is about 30 cm deep. Try not to disturb the sand while adding this water.
4. Allow the system to stand for about 5–10 min and describe what occurs.
   a. Does the pore water appear to mix rapidly with the water above? Why?


   b. If dispersion mixes water masses over time, why is dispersion of the pore water so slow?

5. Using a long-handled spoon or a cup, excavate a deep hole in the sand along one side of the container. The clear side of the container should form one side of the excavation. Transport and deposit the sediment at the opposite side of the container. Try not to create currents or large waves (Figure 12-2).

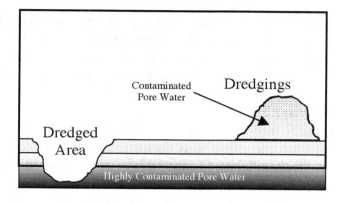

Figure 12-2. Dredging and the release of pore water.

   a. What happens to the pore water as you dig? Why does this occur?

   b. Will this increase the rate of mixing? Why?

   c. Some of the sand that was transported contained pore water. What happened to this water?

6. After digging is finished, allow the system to stand for about 5–10 min and describe any changes that occur.

7. Generate some currents in the water and describe what happens in the area of the excavation and where the dredged sediments were deposited.

## 12-6. BIOACCUMULATION AND BIOMAGNIFICATION

After organisms process food, most of the material rapidly passes through their systems and is returned to the environment in the form of wastes. Metabolic wastes mainly are released in urine and indigestible materials are excreted as feces. Only a small fraction of the matter consumed is retained within their bodies for prolonged periods of time. Yet, some material, especially heavy metals, biocides, and radioactive substances, may be incorporated into skeletal and other body structures and stored for lengthy periods. Because these materials usually occur only in trace amounts in nature, organisms have not evolved processes to rapidly dispose of them when they are abundant. For example, in mammals the body initially stores excess heavy metals in fatty tissues, but gradually disposes of them by incorporating them in growing hair and nails. This is a very slow process, but over time it will remove some portion of these substances. Problems develop when the rate of ingestion exceeds rate of removal. Thus, if an organism repeatedly ingests these pollutants, the material may progressively accumulate within its body. This is **bioaccumulation**, the enrichment of a substance within the body of an organism.

**Biomagnification** refers to the progressive accumulation of a substance in a food chain through bioaccumulation at several trophic levels. For example, phytoplankton, which are microscopic plants, may accumulate small amounts of heavy metals within their bodies (bioaccumulation) because this material is present in the environment. Because zooplankton, which are microscopic herbivores, consume large numbers of phytoplankton, they receive an enriched dose of heavy metals with each meal and have a much greater exposure to the pollutant than one would expect, based on the concentration of the material in the environment. Similarly, each higher level of the food chain potentially becomes enriched in contaminants (biomagnification).

## EXERCISE 6.  BIOACCUMULATION
##                    AND BIOMAGNIFICATION

Procedure

1.  Take several of the small jars labeled meals and one of the trays labeled "organism." The jars are filled with marbles, which represent the atoms present within the meal eaten by an organism. The smaller marbles represent C, H, N and O atoms. The larger marbles represent heavy metals, biocides, or radioactive substances. The holes in the trays are the various pathways by which material consumed by an organism is returned to the environment. All of the holes will allow small marbles to pass through, but only a few holes are large enough to permit the organism to dispose of larger marbles (toxic material).
2.  Feed the animal a meal by pouring the marbles onto the tray. Shake the tray back and forth for a minute over a box. As you shake the tray, many smaller marbles fall through, but only a few of the larger ones are lost.
3.  Continue to feed your animal until all of the meals are gone.
4.  Shake the tray for a minute after the last meal and determine the following numbers:

Small marbles passed through the animal  ＿＿＿＿＿

Small marbles retained by the animal  ＿＿＿＿＿

Large marbles passed through the animal  ＿＿＿＿＿

Large marbles retained by the animal  ＿＿＿＿＿

Percentage of all marbles retained  ＿＿＿＿＿

Percentage of large marbles retained  ＿＿＿＿＿

What process is demonstrated in this part of the exercise?  ＿＿＿＿＿

5.  Each tray bears a number. Trays numbered 1 are zooplankton, trays numbered 2 are small fish and trays numbered 3 are large fish. Zooplankton are eaten by small fish and small fish, in turn, are consumed by large fish. Anyone with a tray numbered 2 can eat any of the trays numbered 1. This is done by pouring the marbles from a tray numbered 1 onto a tray numbered 2. Shake the tray for a minute after each meal. Determine the following numbers after all trays numbered 1 have been consumed.

   Small marbles passed through the animal    _____

   Small marbles retained by the animal    _____

   Large marbles passed through the animal    _____

   Large marbles retained by the animal    _____

   Percentage of all marbles retained    _____

   Percentage of large marbles retained    _____

   What process is demonstrated in this part of the exercise? _____

6.  Repeat this process, but now trays numbered 3 eat trays numbered 2. Determine the following numbers after all trays numbered 2 have been consumed.

   Small marbles passed through the animal    _____

   Small marbles retained by the animal    _____

   Large marbles passed through the animal    _____

   Large marbles retained by the animal    _____

   Percentage of all marbles retained    _____

   Percentage of large marbles retained    _____

   What process is demonstrated by this entire exercise? _____

# Laboratory 13

# Nautical Charts 2 and Piloting

## 13-1. COMPASS READINGS, THE COMPASS, AND MAGNETIC VARIATION

The **North Pole**, also called **Rotational North Pole** or **True North Pole**, is the point on the Earth's surface in the Northern Hemisphere about which the Earth rotates once a day (Figure 13-1). It is one of the two points (antipods) where Earth's imaginary axis intersects the Earth's surface. The graticule, Earth's latitude and longitude grid, are centered about the True North and South Poles. Earth's magnetic field has its own axis and **Magnetic North** and **South Poles**. Because the rotational and magnetic axes are not aligned, rotational and magnetic poles are separated by several miles. Compass needles do not point to the True (Rotational) North Pole, but to the Magnetic North Pole.

A compass is a simple device used to determine the direction to the Magnetic North Pole. It consists of a horizontally pivoted magnetic needle enclosed in a case with a glass top. Inside the case, the needle rotates about a fixed directional scale. Regardless of the direction in which the compass is turned horizontally, the needle always swings so as to point towards the North Magnetic Pole. A **bearing** is the directional angle between the North Magnetic Pole, the compass and some point of interest on Earth's surface (Figure 13-2). Bearings are determined by holding the compass level so that the needle can turn freely, pointing the north end of the compass scale towards the point of interest and reading the angle and direction indicated on the scale by the compass needle when the

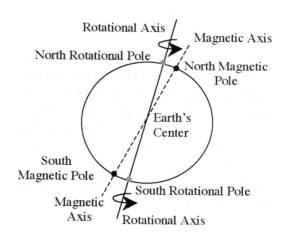

Figure 13-1. Earth's axes. Cross sectional view of the Earth showing the rotational axis (solid line) and rotational poles and the magnetic axis (dashed line) and magnetic poles.

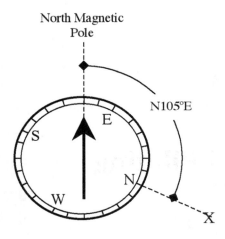

North Magnetic
Pole

N105°E

Figure 13-2. Compass bearing. Point the north end of the compass scale towards the point of interest (X) and read the angle and direction towards which the needle points. Bearing to X is N105°E.

needle ceases to move. On some compasses, East and West directions are reversed on the scale so that directions can be read directly (Figure 13-2).

**Magnetic variation** or **declination** is the angle between True North, a point on Earth's surface and Magnetic North. The amount of magnetic variation changes with location across Earth's surface because the angular relationship between the Magnetic Pole and Rotational Pole varies with location (Figure 13-3). Variation is 0° along the **agonic line**, an imaginary line that intersects both the Magnetic and Rotational poles. To the east and west of this line, variation increases and reaches a maximum of 180° for points located between the North Rotational and North Magnetic poles in the Northern Hemisphere. Here, compass needles point southward, not north. Because of irregularities within Earth's magnetic field, the agonic line is highly irregular and does not form a great circle.

Changes within Earth's magnetic field cause the magnetic poles to slowly move and the position of the agonic line to shift westward with time. Presently, part of the agonic line passes through the center of the U.S. near the Mississippi River. As the magnetic pole migrates, the variation at any location slowly changes. For example, if the variation in 2000 at some point was 21°45′ and the annual change is −5′/yr, in 2007 the cumulative annual change would be $07\text{yr} \times -5'/\text{yr} = -35'$ and the variation is now $21°45' - 35' = 21°10'$.

Bearings can be expressed in a variety of numerical and directional systems, including azimuths, quadrant compass bearings and compass points. **Azimuths** are bearings measured clockwise from the north, from 0° to 360°. For example, 90° east of north is 90°, and 90° west of north is 270°. Both 0° and 360° are Due North. For azimuths, only degrees are stated, not directions.

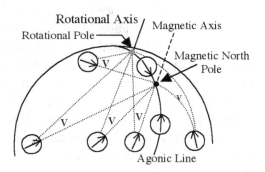

Figure 13-3. Changes in variation with location. Variation is 0° on the agonic line and reaches 180° for points between the North Magnetic and North Rotational poles where the needle points southward. V = variation.

**Quadrant compass bearings** are directional compass angles measured eastward or westward from either north or south, whichever is closer. In designating a compass bearing, both direction and degrees are used. The maximum bearing in this system is 90°. Thus, **Due North** and **Due South** are 0°N and 0°S, respectively. N/S90°E is **Due East** and N/S90°W is **Due West**. A point 45° clockwise from Due North is N45°E, and 45° counterclockwise is N45°W. A point 45° clockwise from Due South is S45°W, and 45° counterclockwise is S45°E.

**Compass points** are 32 named compass directions. There are four **cardinal points-**north, east, south, and west-which divide the compass into four quadrants. These quadrants are then equally divided by four secondary points (NE, SE, SW, and NW), eight tertiary points (NNE = north by northeast, ENE = east by northeast, ESE, SSE, etc.), and finally by 16 additional points (N by E = north by east, NE by N = northeast by north, etc). Compass points are interesting and of historical significance, but they are not very accurate. Most often the direction of interest does not fall exactly on any of the 32 points.

A major problem with magnetic bearings is that changes in Earth's magnetic field require them to be frequently updated to remain accurate. Bearings relative to the rotational pole never change and they can be easily plotted on the graticule. Thus, it is common to convert magnetic bearings into what are called **True bearings**, bearings related to the North Rotational Pole and reported as **degrees True** (°T). To do this, it is necessary to adjust for the variation for the time when the magnetic bearing was taken. This is most easily done using azimuths, but can be accomplished using any bearing system. Always be certain as to the type of bearings, True or magnetic, that one is using.

Areas east of the agonic line have a variation to the west and compass needles are pointing west of True North (Figure 13-4A). Resulting bearings are too large. Uncorrected compass readings are made True (corrected) by subtracting the variation from the reading. (*Subtract variations to the west.*) Areas west of the agonic line have variation to the east and compass needles are pointing east of True North (Figure 13-4B). Resulting bearings are too small. Uncorrected readings are made **True** (corrected) by adding the variation to the original reading. (*Add variation to the east.*)

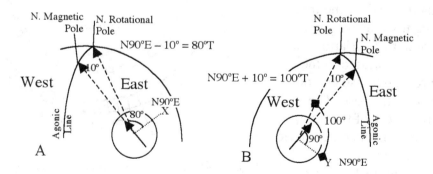

Figure 13-4.  Correcting for variation east (A) and west (B) of the agonic line.

Example 1. A compass reading towards some point X (Figure 13-4A) is N90°E (azimuth 90°) for an area with a magnetic variation of 10°W. Because the declination is to the west, the location of the compass must be east of the agonic line. This means the compass needle points 10° farther to the west than it should relative to the location of the North Rotational Pole. Correct the reading for True North by subtracting 10° from the compass reading: 90° − 10° = 80° T (azimuth).

Example 2. A compass reading towards some point Y (Figure 13-4B) is N90°E (azimuth 90°) for an area with a magnetic variation of 10°E. Because the declination is to the east, the location of the compass must be west of the agonic line. This means the compass needle points 10° farther to the east than it should relative to the location of the North Rotational Pole. To correct the reading for True North, add 10° to the compass reading: 90° + 10° = 100° T.

North, south, east, and west on a chart can be determined by examining one of the chart's compass roses. There are usually four roses on a large chart, one in each corner. A **compass rose** usually consists of two concentric circles, each divided into 360° (Figure 13-5), so bearings are measured as azimuths. The outer circle is aligned to the rotational poles and the graticule. The inner circle is aligned to the location of the magnetic pole at the time the chart was made. On the outer circle, **Due North (True North)** is indicated by a five-pointed star at 0°. The star points towards Earth's North Rotational Pole. **Due East** is 90° to the right of north, **Due South** is 180° to the right of north and **Due West** is 270° to the right of north. On most charts, north is located at the top, but one should never assume this. Always check chart orientation.

The inner circle of a compass rose is aligned to the Magnetic North Pole (Figure 13-5). It shows the direction in which compass needles pointed at the time the chart was made. An arrow extends from 0° on the inner circle, pointing towards magnetic north. The amount of variation for the chart is usually printed in the center of the rose, but also can be read directly from the rose. The magnetic north arrow of the inner circle points towards the outer circle of the rose at the value of the declination. On Figure 13-5, the arrow points between 25° and 30° and the declination printed in

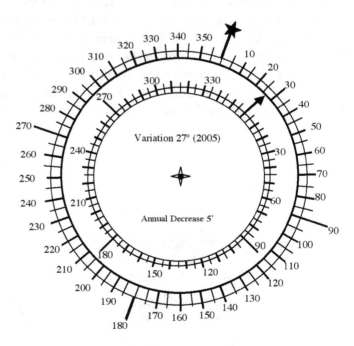

Figure 13-5. Compass rose. Outer scale is aligned to True North and inner scale to Magnetic North. Center of rose provides variation at the time the map was made and annual change in variation.

the center of the rose is 28°. The center of the rose also states the year the variation was determined and the amount that variation changes annually. Compass roses and annual variation must be updated every 20 years to remain accurate because the annual variation change is not constant over long periods of time.

A second type of compass rose that one occasionally sees on older charts divides the rose into compass points.

## EXERCISE 1.   COMPASS ROSES, BEARINGS AND MAGNETIC VARIATION

1. On the compass rose illustrated in Figure 13-6, label the compass points listed below: N, S, E, W, NE, SE, SW, NW, NNE, ENE, ESE, SSE, SSW, WSW, WNW, NNW, N by E, NE by N, NE by E, E by N, E by S, SE by E, SE by S, S by E, S by W, SW by S, SW by W, W by S, W by N, NW by W, NW by N, N by W.

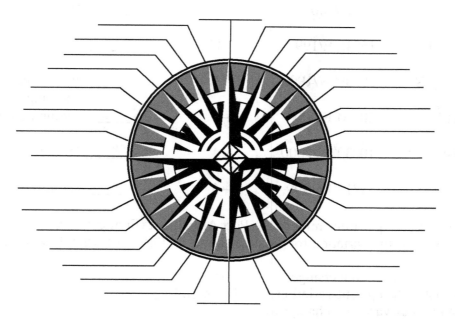

Figure 13-6.  Compass rose illustrating 32 compass points.

2. How many degrees separate any two adjacent points on the compass rose? _____

3. Give the azimuths and quadrant bearings for each of the following compass points:

| Point | Azimuth | Quadrant Bearing | Point | Azimuth | Quadrant Bearing |
|-------|---------|------------------|-------|---------|------------------|
| NE | _____ | _____ | NNE | _____ | _____ |
| N by E | _____ | _____ | S | _____ | _____ |
| NE by N | _____ | _____ | NE by E | _____ | _____ |
| E by N | _____ | _____ | E | _____ | _____ |

4. What is the relationship between N60°E and S60°W?

5. What are the advantages of azimuths over quadrant bearings and compass points?

6. Below are a series of uncorrected compass readings, declinations and annual corrections. Determine the corrected compass reading for 2007 as requested.

| Compass Reading | Declination/ for year | Annual Correction | Corrected Reading |
|-----------------|----------------------|-------------------|-------------------|
| a. N67°E | 23°W/2001 | +6′ | _____°T |
| b. 199°T | 15°45′E/1999 | −8′ | _____ (compass point) |
| c. W by N | 19°09′W/1999 | +12′ | _____ (quadrant compass bearing) |
| d. N19°W | 20°50′E/2004 | −5′ | _____ (compass point) |
| e. NW by N | 10°30′W/2000 | +4′ | _____°T |
| f. 175°T | 11°13′E/1999 | −6′ | _____ (compass bearing) |

7. The following questions refer to the Port Arbit chart (Figure 13-7).
   a. What was the variation for the area shown on this chart when first printed?
   _____

   b. What is the annual change in variation? _____
   c. What is the current variation for this region? _____
   d. Why does variation change with time?


   e. If variation is decreasing west of the agonic line as the pole moves westward, does that mean it must be increasing east of the agonic line? Why?


   f. What is the latitude and longitude of Point Maud on Essen Island?
   _____

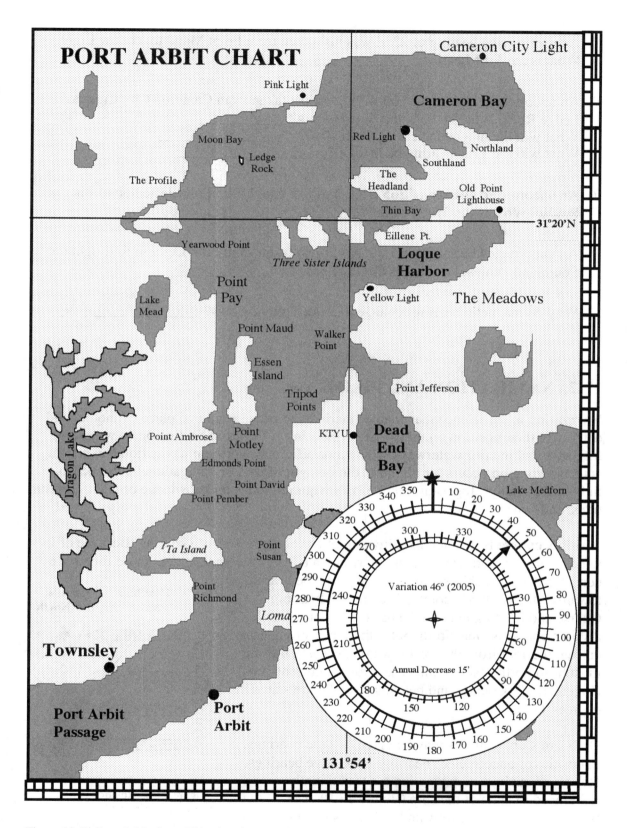

Figure 13-7. Port Arbit chart. This chart is not to be used for navigation.

8.  An uncorrected bearing of 23°E was just taken from Point Maud to Yellow Light. What is the True bearing? _____

9.  An uncorrected bearing of 200°42′E was just taken from Cameron City Light to Ledge Rock. What is the is the True bearing? _____
    What will the magnetic bearing be in 2010? _____
    What will the True bearing be in 2010? _____

10. An uncorrected bearing of 265°31′E was just taken from Point Jefferson to The Profile. What is the True bearing? _____

11. An uncorrected bearing of 65°39′E was just taken from Pink Light to Red light on Southland Peninsula. What is the True bearing? _____

12. Why do magnetic bearings change, but not True bearings? _____

## 13-2.  NAVIGATION AND PILOTING

**Navigation** is the science that enables mariners to travel from one place to another and determine their position on the open sea. **Piloting** is the art of finding one's way along the shore and interior waterways. The major difference between navigation and piloting is that navigation occurs far offshore, out of sight of land or navigational aids such as lighthouses, buoys, foghorns, and bells. Navigation relies on knowledge of stars, currents, tides, and waves.

A **ship's course** is its intended direction of travel expressed as a bearing. Because of winds, currents, and waves, a ship may deviate from its course, unless these forces are considered and the ship's direction adjusted. The bearing towards which a ship is pointed is called its **heading**. Course and heading may not be the same. For example, a ship's course is Due North, but if there is a strong east current (current from the west towards the east), the ship must head northwest to compensate for the current (Figure 13-8). Courses and headings usually are stated in degrees True.

On board ship, bearings may be given as True bearings, magnetic (or uncorrected) bearings or relative bearings. **Relative bearings** in navigation are traditionally referenced to parts of the ship (Figure 13-9). The front of the ship is the **bow** and directions toward the bow are called **forward**. The direction directly in front

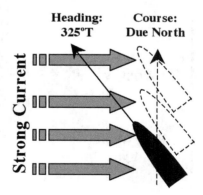

Figure 13-8. Relationship of course, bearing and an eastward current. Outline of ship shows progression with time.

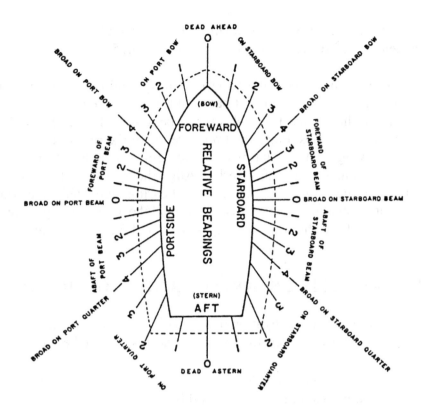

Figure 13-9. Relative bearings on a ship. Points are 11.25° apart.

of the bow is **dead ahead**. The rear of the ship is the **stern** and directions toward the stern are called **aft** (also **abaft**). Directly behind the stern is **dead astern.** As one faces forward on a ship, **starboard** is to the right and **port** is to the left. Other directions on board ship are shown in Figure 13-9.

Relative bearings are frequently stated in terms of points. A **point** is equal to 11.25°. For example, one point on the port bow would be 11.25° to the left of dead ahead. Two points on the starboard quarter is 22.5° to the right of dead astern. Directions at 45°, 90° and 135° on portside are **broad on port bow, broad on port beam,** and **broad on port quarter**, respectively. Similar names apply to the starboard. The 32 points of relative bearing on the ship are similar divisions to the 32 points on a compass rose.

Navigators and pilots must be able to convert relative bearings to True bearings. For example, if a ship is headed 10°T and a lighthouse appears three points abaft of the starboard beam, its True bearing is 133.75°T (Figure 13-10) [90° from dead ahead to

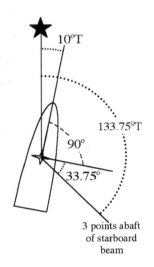

Figure 13-10. Converting relative bearings to True bearings. Object located 3 points abaft of starboard beam is at 133.75°T.

broad on starboard beam plus three points abaft of the starboard beam (11.25° × 3 = 33.75°) and the heading of 10°T. 90° + 33.75° + 10° = 133.75°T].

An easy way to convert relative bearings to True bearings and vice versa is to use a transparent True compass rose. On the last pages of this book are two sheets containing several useful tools for these labs. These sheets can be removed from the book and photocopied onto transparencies. Place the transparency of the rose over the printed image of the ship showing relative bearings (Figure 13-9). Be certain that the cross hairs (center points) of the compass and ship overlap. Align the rose so that the bow points towards the heading of the ship. Locate the relative bearing on the ship and where a line from this bearing intersects the compass rose, then read the True bearing. Alternatively, place the transparency of the ship over a compass rose and follow the same procedure.

## EXERCISE 2.  RELATIVE BEARINGS AND TRUE BEARINGS

1.  Relative bearings to various navigational aids and coastal features and the ship's heading are given below. Convert the relative bearings to True bearings.

| Heading | Object Location | True Bearings |
|---------|-----------------|---------------|
| 0°T | Buoy, one point on starboard bow | _____ |
| 30°T | Buoy, one point on port bow | _____ |
| 0°T | Buoy, one point on port quarter | _____ |
| 30°T | Buoy, one point on starboard quarter | _____ |
| 40°T | Light, one point forward of starboard beam | _____ |
| 70°T | Whistle buoy, three points of starboard quarter | _____ |
| 62°T | Reef, broad on port beam | _____ |
| 225°T | Flashing light, broad on port bow | _____ |
| 15°T | Rocks, two points abaft of port beam | _____ |
| 180°T | Groaner buoy, dead ahead | _____ |
| 342°T | Wreck buoy, two points forward of port beam | _____ |
| 235°T | Breakers, one point on port bow | _____ |

2. True bearings to various aids and coastal features are determined using charts and the ship's location. On board ship these must be translated into relative bearings, which in part are determined by the ship's heading.

| Heading | True Bearings | Relative Bearings |
|---------|---------------|-------------------|
| 0°T | 90°T | _____ |
| 30°T | 120°T | _____ |
| 0°T | 225°T | _____ |
| 30°T | 255°T | _____ |
| 54°T | 349°T | _____ |
| 300°T | 95°T | _____ |
| 180°T | 0°T | _____ |
| 90°T | 232.5°T | _____ |
| 156°T | 291°T | _____ |
| 263°T | 180°T | _____ |

## 13-3. CHARTING A COURSE WITH PARALLEL RULES AND A COMPASS ROSE

Before traveling by ship to a destination, it is customary to **chart a course**. This means to plot on a chart your planned path from point of origin to destination and to determine the distances to be traveled at a given bearing. This is most commonly done with parallel rules. **Parallel rules** are two straight edges hinged together so that they remain parallel as they move. The following steps are taken to chart a course:

Procedure

1. Identify your present location and destination.
2. Join the two points with a series of connecting straight lines, called **legs**. Obviously, when traveling by boat, the course can not cross land. Although a curved course may be shorter, it is not possible to express a curve as a course because, in a curve, direction changes continuously and there is no way to indicate this with bearings.
3. Number each leg and measure and record their length in nautical miles.
4. Using parallel rules.
   a. With the rules together, place one edge of the parallel rules along the line representing a leg of the course (Figure 13-11).
   b. Locate the nearest compass rose.

c. Hold the rule that is farthest from the compass rose firmly in place and slide the other rule toward the rose. Hold the rule closer to the rose firmly in place and slide the other rule towards it. Continue alternating the rules and slowly "walk" them towards the compass rose. As long as the rules have not slipped, they will still be parallel to the course leg. When one rule cuts through the center of the rose and the outer scale, read the direction from the outer circle for degrees True in which the ship must travel. Because the rule cuts through the scale in two places, be certain to choose the bearing for the direction in which the ship will be traveling.

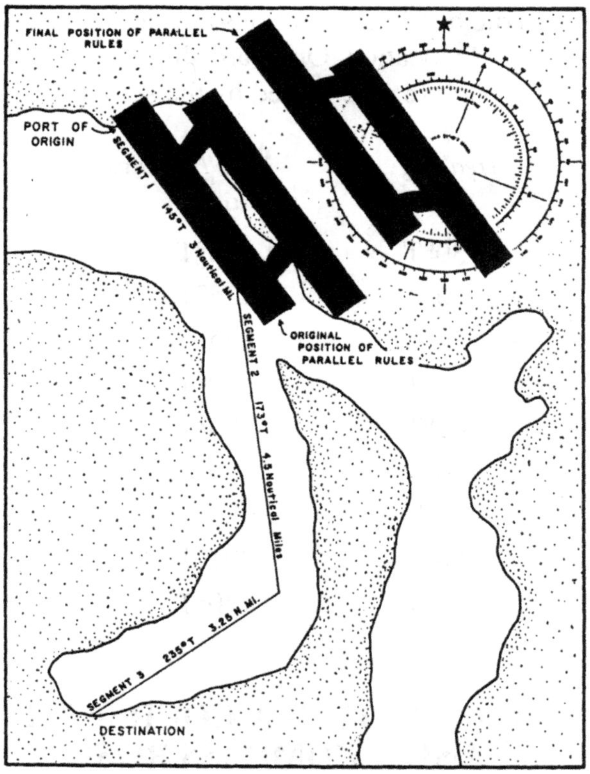

5. Determine and record the bearing for each leg of the trip. Whenever taking or plotting bearings, it is important to be precise. A difference of only one degree projected over several miles can result in a large deviation between destination and plotted course. Errors also tend to be cumulative. Each additional leg of the trip plotted will be farther off course because of the initial error.

Figure 13-11. Plotting a course with parallel rules. "Walk" the rules from the course to the rose and read the bearing in the direction to be traveled.

## EXERCISE 3.  CHARTING AND FOLLOWING A COURSE

1. A course has been charted from Port Arbit to various locations where supplies are to be delivered (Figure 13-12). Give the bearings and distances to be traveled in the trip to and from Port Arbit. The outward trip is shown as solid lines (1–7) and the return trip as dashed lines (8–10).

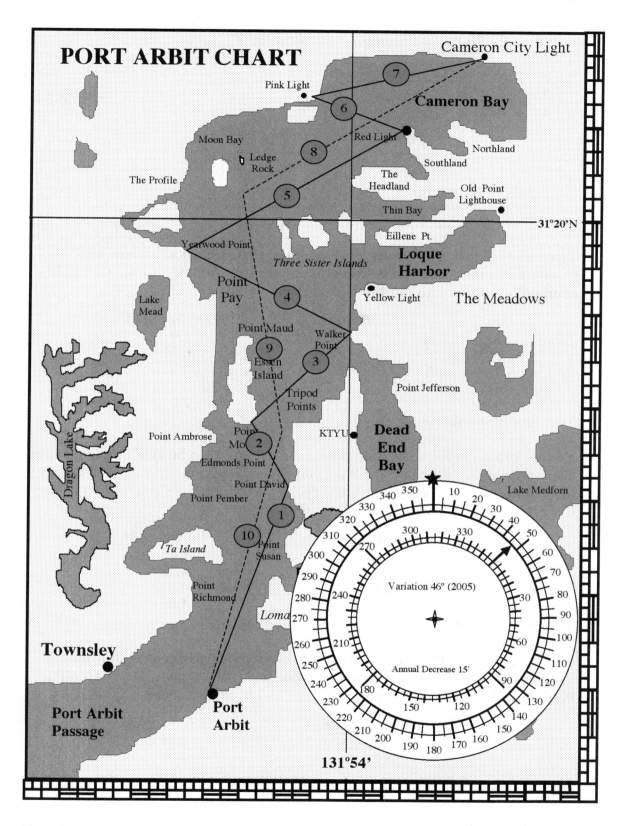

Figure 13-12. Port Arbit chart. Solid lines (1–7) are outward legs of trip; dashed lines (8–10) inbound.

| Leg | Magnetic Bearings | Bearing °T | Distance NM |
|-----|-------------------|------------|-------------|
| 1 | _____ | _____ | _____ |
| 2 | _____ | _____ | _____ |
| 3 | _____ | _____ | _____ |
| 4 | _____ | _____ | _____ |
| 5 | _____ | _____ | _____ |
| 6 | _____ | _____ | _____ |
| 7 | _____ | _____ | _____ |
| 8 | _____ | _____ | _____ |
| 9 | _____ | _____ | _____ |
| 10 | _____ | _____ | _____ |

2.  You have been invited to a party. No location is given, but a detailed course is provided. Plot the course and determine where the party will be located. Your point of origin is Townsley. Use Figure 13-7 to chart the course.

| Leg | Bearing °T | Distance NM |
|-----|-----------|-------------|
| 1 | 090 | 1.9 |
| 2 | 018 | 6.5 |
| 3 | 350 | 4.0 |
| 4 | 057 | 3.6 |
| 5 | 135 | 2.0 |

Your final destination is_____

3.  You are looking to buy some waterfront real estate. An ad in Sunday's paper provides the instructions listed below to visit the property, which is not serviced by any roads. Convert the magnetic bearings to True bearings. Your origin is Cameron City Light. Use Figure 13-7 to plot the course.

| Leg | Magnetic Bearings | °T | Distance NM |
|-----|-------------------|-----|-------------|
| 1 | 187 | _____ | 5.65 |
| 2 | 125 | _____ | 4.25 |
| 3 | 145 | _____ | 4.0 |
| 4 | 205 | _____ | 1.0 |

Your final destination is_____

4  You are a welder in Port Arbit who specializes in repairing large steel structures. A work order sends you to the following location to do some repairs. Convert the True bearings to True compass points. Use Figure 13-7 to plot the course.

| Leg | True Compass Points | °T | Distance NM |
|-----|---------------------|-----|-------------|
| 1 | _____ | 15 | 3.85 |
| 2 | _____ | 09 | 4.35 |
| 3 | _____ | 120 | 2.00 |
| 4 | _____ | 156 | 1.25 |
| 5 | _____ | 260 | 0.75 |

Your final destination is _____

## 13-4.  CHARTING A COURSE WITH A PROTRACTOR OR TRANSPARENT COMPASS ROSE

If a chart does not have a compass rose or you do not have parallel rules, it is still possible to determine directions in degrees True by using the intersection of course legs with lines of latitude and longitude. A compass rose is aligned so that a line passing through 0°T and 180°T is a north-south line, a line of longitude. A line passing through 90°T and 270°T on the rose is an east-west line, a line of latitude. Thus, a leg that intersects a line of longitude inscribes the same angle as would the line if it passed through the center of a compass rose. Using a protractor (Figure 13-13), the angle between the line of longitude and the leg can be measured and the True bearing read directly. If the leg does not intersect a line of longitude, the leg can be projected across the chart until it does. If this is not possible, bearing can be determined using lines of latitude (Figure 13-13). Lines of latitude are perpendicular to lines of longitude on nautical charts. At the point of intersection between the leg and the line of latitude, draw a line perpendicular to the line of latitude. This line is a line of longitude and it can be used to determine the bearing of the leg.

Alternatively, the course can be plotted using a transparent compass rose. Place the rose so that its cross hairs sit atop the point of intersection of the leg, or its extension, with a line of latitude or longitude. For a line of longitude, rotate the rose until the 0° and 180° marks lie upon the line of longitude. For a line of latitude, rotate the rose until the 270° and 90° marks lie upon the line of latitude. The rose is now aligned with True North. Read the bearing in the direction in which the ship will travel.

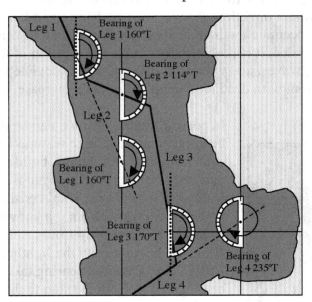

Figure 13-13.  Using a protractor to determine bearings in degrees True. Dashed lines are projection of legs. Dotted lines are lines of longitude drawn perpendicular to latitude where crossed by a leg of the course.

## EXERCISE 4.  USING A PROTRACTOR OR TRANSPARENT COMPASS ROSE TO DETERMINE BEARING

On Figure 13-14, Calmeton chart, two courses have been charted. Determine the bearing and distance for each leg of the courses.

Course 1

| Leg | Bearing °T | Distance NM |
|-----|-----------|-------------|
| 1 | _____ | _____ |
| 2 | _____ | _____ |
| 3 | _____ | _____ |
| 4 | _____ | _____ |

Course 2

| Leg | Bearing °T | Distance NM |
|-----|-----------|-------------|
| A | _____ | _____ |
| B | _____ | _____ |
| C | _____ | _____ |
| D | _____ | _____ |
| E | _____ | _____ |
| F | _____ | _____ |

## 13-5.  DETERMINING POSITION

As a ship travels, it is necessary to verify that it is **on course**, located where it should be and traveling in the correct direction. Position can be determined in several ways. **Dead reckoning** is the least accurate method. Knowing a ship's point of origin, speed, direction, and time traveled, it is possible to estimate the location of the ship. However, the effects of currents, waves, and winds can cause a ship to deviate from its course. With time, inaccuracy inherent with dead reckoning will greatly increase.

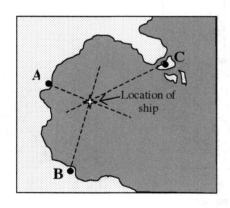

Figure 13-15.  Determining position using bearings to three features A, B and C.

In coastal waters, position can be determined by taking bearings to at least three features of known location, such as lighthouses, buoys, buildings, etc. A True or magnetic bearing is taken from the ship to each of the features. To locate the ship's position, it is necessary to work backwards from the known locations. For example, if the bearing from a ship to a lighthouse is 30°T, the bearing from the lighthouse to the ship is 210°T, 180° from the bearing taken. Either convert bearings taken from the ship to the locations into bearings from the known locations back to the ship, or when plotting the bearings, draw the lines in the opposite direction to the bearings. The intersection of the three lines from known locations indicates the location of the ship (Figure 13-15).

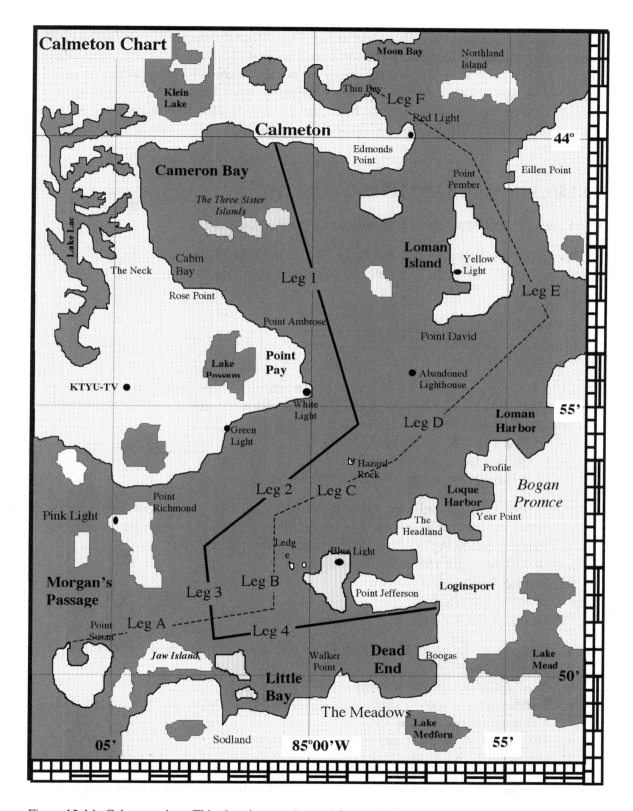

Figure 13-14. Calmeton chart. This chart is not to be used for navigation.

If parallel rules are not available to transfer bearings from the compass rose on the chart to the points of known location, a protractor or transparent compass rose can be used. Draw a line of latitude or longitude through the point of known location using the technique shown in Laboratory 9, Section 2. Place the cross hairs of the compass rose over the known location and align the rose with the line of latitude or longitude, as in section 13-6. Locate the bearing and draw a line in the appropriate direction to represent the bearing.

## EXERCISE 5.  DETERMINING POSITION

Use Figure 13-16, Port Arbit chart, for the following questions.

1. As part of a scavenger hunt, you bury a prize on Point Maud, the northern tip of Essen Island. For a clue to locate the prize, you give True bearings to the following locations:
   a.  Pink Light         _____
   b.  Yellow Light       _____
   c.  KTYU Antenna       _____
   d.  Ledge Rock         _____

2. In 2005, you have found a good fishing area near the Three Sisters Islands. So as to be able to return to it later, you located your position by taking the bearings listed below. Mark your position on the chart.
   a.  To Pink Light 335° (magnetic)
   b.  To Yellow Light 60° (magnetic)
   c.  To middle of Tripod Points, at 104° (magnetic)

3. While boating, your outboard motor falls overboard. You plan to row ashore, get your scuba gear and return to salvage the motor. So as to precisely locate your position, you take the following bearings. Mark your position on the chart.
   a.  To Yellow Light 158°T
   b.  To Pink Light 350°T
   c.  To Red Light 78°T
   d.  To Ledge Rock 260°T

4. You are new to the area and are lost. From your boat you take bearings to the following locations:
   a.  A tall radio or television tower at 36°T
   b.  Two towns on opposite sides of the channel, one due south and the other at 252°T
   c.  A distant yellow light at 25°T
   d.  The eastern tip of an island about 1.5 nm away at 9°T
       Identify your position on the chart.

5. If there was not a compass rose on the chart and only magnetic bearings were provided, would it be possible to determine your position using a protractor or transparent compass rose?

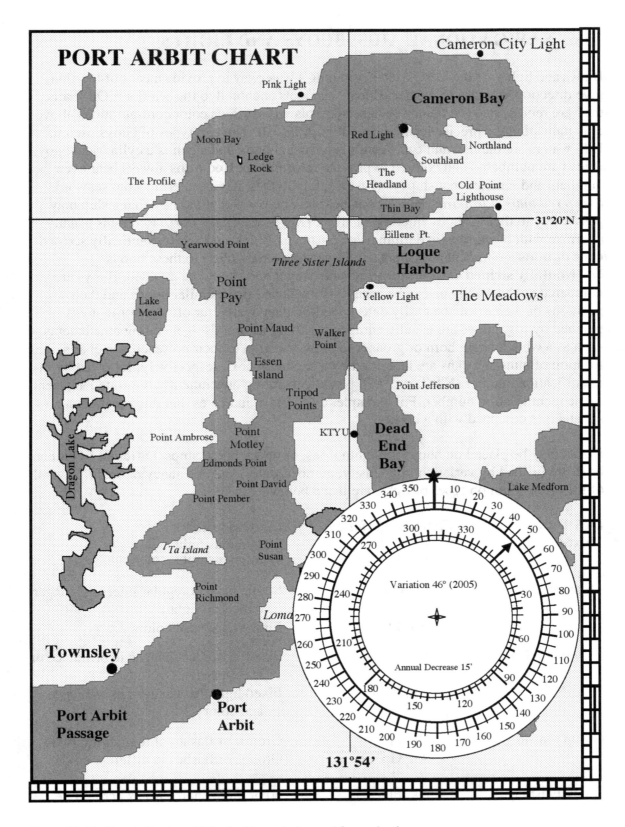

Figure 13-16.  Port Arbit chart. This chart is not to be used for navigation.

## 13-6. NAVIGATIONAL AIDS: BUOYS AND LIGHTS

Buoys and lights are devices placed in or along a waterway to provide means of location, give direction, or warn of danger. **Buoys** are floats anchored to the sea floor. On charts they are represented by diamond-shape symbols with various color configurations above dots indicating the precise location (see Figure 13-19). The two types of buoys used in U.S. waters are the nun and can. A **nun** is a conical buoy and a **can** is a cylindrical buoy. Buoys are numbered and colored to provide information. **Red nuns** always bear even numbers and indicate that ships may pass safely to their left as entering from seaward. *(Keep red nuns starboard.)* **Black cans** bear odd numbers and indicate that a ship may pass safely to their right as entering a harbor. *(Keep black cans port.)* Red nuns usually alternate with black cans in defining safe channels. **Black and white, vertically striped cans** or **nuns** are midchannel markers. Ships should pass close to these buoys. **Horizontally striped cans** or **nuns** indicate danger and should be avoided. Buoys are numbered so that each one can be easily distinguished from similar buoys and their positions on a chart can be easily determined. In deep water, far offshore or in areas plagued by fog, buoys are usually equipped with lights or noise-generating attachments, such as a whistle, bell, horn or groaner to draw attention to their presence. On charts, the location of nuns and cans are identified by a capital N or C, respectively, followed by the identification number (N"#" or C"#"). Noise-generating attachments are printed adjacent to the identification number. For example, *C"6" Horn* indicates that this is can number six and it is equipped with a horn.

Lights may be placed on buoys, in lighthouses, or on anchored ships. Different lights can be distinguished by variation in intensity, number, color, and constancy of beam. Several of the more common types of lights are listed below.

| Light Type | Chart Abbreviation | Characteristic |
|---|---|---|
| Fixed | F. | Continuous light |
| | Alt. F | Continuous light, color changes |
| Flashing | Fl. Rot | Single flash at regular intervals, longer dark than light |
| | Alt. Fl | Flashes different colors |
| Fixed and Flashing | F. Fl | Fixed light that changes in brilliance or is eclipsed |
| | Alt. F. Fl | Fixed light that changes in brilliance and color |
| Group Flashing | Gp. Fl | Groups of flashes at regular intervals |
| | Alt. Gp. Fl | Groups of flashes in different colors |
| Quick Flashing | Qk. Fl | 60 or more short flashes per minute |

| Interrupted Quick Flashing | I. Qk. Fl | Quick flashes for 4 sec then 4 sec darkness |
| --- | --- | --- |
| Short-Long Flashing | S-L. Fl | Short flash of 0.4 sec, long flash of 3.2 sec; repeated 6–8 times a minute |
| Occulting | Occ. | Steady light totally eclipsed at intervals; period of light greater than or equal to darkness |

Period of flashing, color, height of light, and distance the light is visible are printed on the chart beside the location of the light. Height is provided because it largely determines the distance over which the light will be visible. The higher the light, the farther it can be seen. Abbreviations of colors and several examples of light descriptions are provided below.

| Light Color | Abbreviation | Light Color | Abbreviation |
| --- | --- | --- | --- |
| Violet | Vi | Blue | Bu |
| Green | G | Orange | Or |
| Red | R | White | W |
| Amber | Am | | |

| Chart Abbreviation | Explanation |
| --- | --- |
| FW | Fixed white light, continuous beam |
| FG | Fixed green light, continuous beam |
| FBu | Fixed blue light, continuous beam |
| Fl R 5sec | Red light flashes every 5 sec |
| Fl G 3sec 28ft 9M | Green light flashes every 3 sec, height of 28 ft, seen for 9 nm |
| Alt Fl W & Vi 10sec | Alternating flashes of white and violet light every 10 sec |
| F Fl R 5sec | Fixed and flashing red light; continuous beam of red brightens every 5 sec |
| FW Alt Fl R 40sec | Fixed white alternating with flashes of red; continuous beam of white light interrupted by red flash each 40 sec |
| Gp Fl W(3)20sec | Group flashing white; three white flashes every 20 sec |
| Gp Fl W (143) 30sec | Group flashing white; one white flash, four white flashes, three white flashes, eclipse (darkness), and repeated every 30 sec |
| Occ W 10sec | Occulting white; continuous white beam broken by short eclipse (darkness). Total time of light and darkness is 10 sec |

## EXERCISE 6.  NAVIGATIONAL AIDS

1.  Explain what each abbreviation means:
    a.  FI 6sec Horn _____
    b.  FI R 4sec _____
    c.  Fl 5sec 39ft 15M _____
    d.  Alt FI R&W 20sec 27ft 16M Horn _____
    e.  FG 22ft 5M Bell _____
    f.  Occ R 20sec Groaner _____
    g.  Qk FI 12Ft 5M _____
    h.  Rot W&Y _____
    i.  Gp Fl (3) 10sec 27Ft 19M Horn _____
    j.  N"6" R _____
    k.  C"7" B _____

2.  While boating in Lacy Sound near Kjolen Islands (Figure 13-17) on a foggy night, you become lost. Because there are hazards in some areas around the islands, it is essential for you to know your location. From your boat you can see and hear several navigational aids. The bearings you have taken are listed below. On the chart, identify your position.
    a.  92°T yellow light that flashes about two times a minute.
    b.  240°T orange light that flashes each second.
    c.  173°T four lights that flash every 20 sec.
    d.  6°T violet light that briefly fades to darkness.
    e.  Foghorn heard to the southeast.

3.  While fishing one night, a thick patchy fog settles on the water. You know that you had been somewhere south of Nargaport (Figure 13-17), but you have been drifting for several hours and the tidal current has been strong because it is the night of the new moon. To determine your position, you have taken the bearings listed below. On the chart identify your position.
    a.  Foghorn some distance to the southwest.
    b.  304°T alternating short and longer flashes of white light repeating several times a minute.
    c.  218°T faint white light that flashes every 15 sec barely visible in the fog.
    d.  63°T bright green light that appears every few seconds.

    You plan to dock and spend the night in Ivers City. What is the bearing and distance to your destination? _____

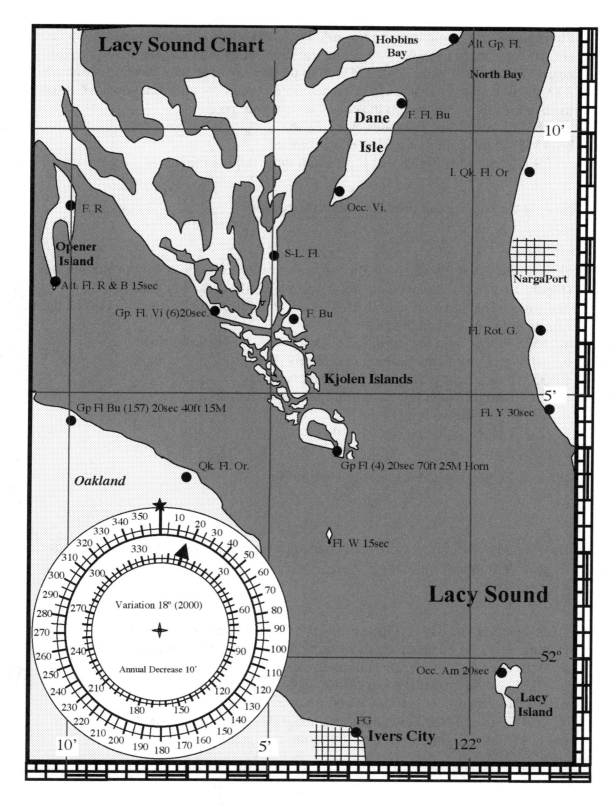

Figure 13-17.  Lacy Sound. This chart is not to be used for navigation.

Shortly after you begin heading towards Ivers City, the fog becomes extremely thick and you are forced to travel slowly and use dead reckoning. As you travel you can hear a foghorn first to the west and then to the northwest of your location. After about 3 hr, the fog again begins to clear. To determine your new position, you have taken the bearings listed below. On the chart identify your position.

   a.   185°T amber light that fades every 20 sec.
   b.   15°T yellow light that flashes every 30 sec.
   c.   290°T four white lights that flash about three times a minute.

What is the bearing and distance to Ivers City now? _____

4.  On a bright, starry, moonless night you are traveling from Ivers City to Hobbin Bay (Figure 13-17). Because you are not familiar with the area, you must rely on navigational aids to pilot your boat. You take the following bearings to various lights. Locate your position on the chart.

   a.   65°T orange light that flashes for 4 sec and then is dark for 4 sec.
   b.   10°T group of lights flashing in different colors.
   c.   352°T blue light that varies in brightness.
   d.   291°T bright violet light that briefly fades to darkness.
   e.   231°T continuous bright blue light.
   f.   132°T green light that briefly appears.

What is the bearing and distance to Hobbin Bay?

Does having bearings to more than three known points assist in locating your position or make it more difficult? Why?

5.  After spending the day on Opener Island, you are returning to Ivers City (Figure 13-17). You take the following bearings. Mark your position on the chart.

   a.   310°T red and blue lights alternately flash four times a minute.
   b.   52°T several violet lights that flash about 18 times a minute.
   c.   160°T orange light that flashes 60 times a minute.
   d.   234°T several blue lights that flash about three times a minute.
   e.   138°T white light that flashes four times a minute.

6. You are a captain of a freighter who has cargo to deliver in the towns along the shore of Wasser Harbor (Figure 13-18). You are unfamiliar with the waters of Wasser Harbor and must rely on the various navigational aids to locate the only safe deep-water channel. On the chart provided, draw the course you will follow to make deliveries to Port Allen, Monroe City, and Stafferton.

a. Will nuns be located on the starboard or port side of the ship as it returns to sea?

b. Why are nuns and cans numbered?

c. Why are buoys used to indicate the middle of a channel?

7. What are the advantages of visual aids, such as lights, over audio aids, such as horns, bells, and groaners, in locating one's position? Under what conditions is each superior?

## 13-7. NAUTICAL CHART SYMBOLS AND ABBREVIATIONS

Numerous symbols and abbreviations appear on nautical charts. Several of these were listed in Laboratory 4, Marine Sedimentation. Additional abbreviations and symbols are provided in Figure 13-19.

## EXERCISE 7.  USE OF CHART SYMBOLS

You are to pilot a ship from the ocean through the various channels to the town of Home Port. Currently, the ship is just off the west coast by a large coral reef in the northwest corner of the chart (Figure 13-20). Before entering the channel, you must find the safest and shortest course. Many hazards block your passage, but all of these are shown on the chart. Because your ship "draws" 8 ft of water (extends below the surface 8 ft), you hesitate to venture into water shallower than 9 ft because you are uncertain about the tide. As you pass each symbol on the chart, label what it indicates.

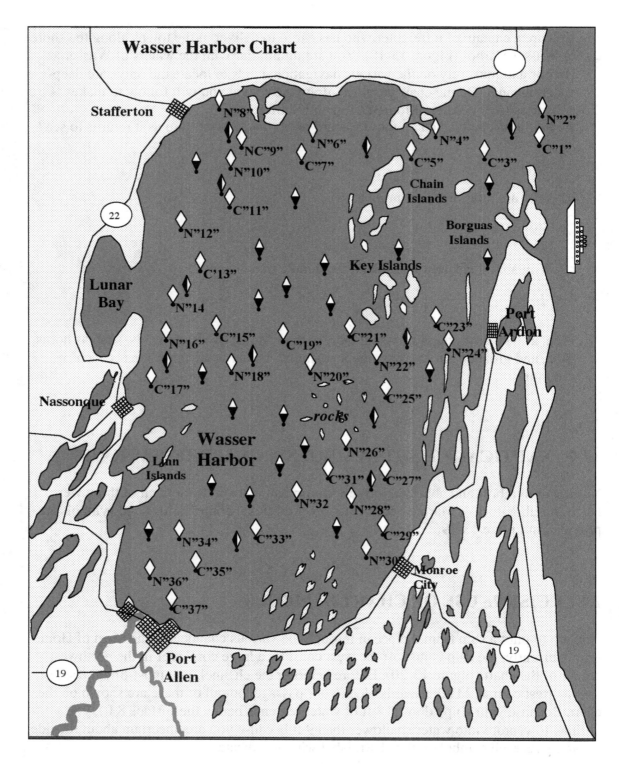

Figure 13-18.  Wasser Harbor chart. This chart is not to be used for navigation.

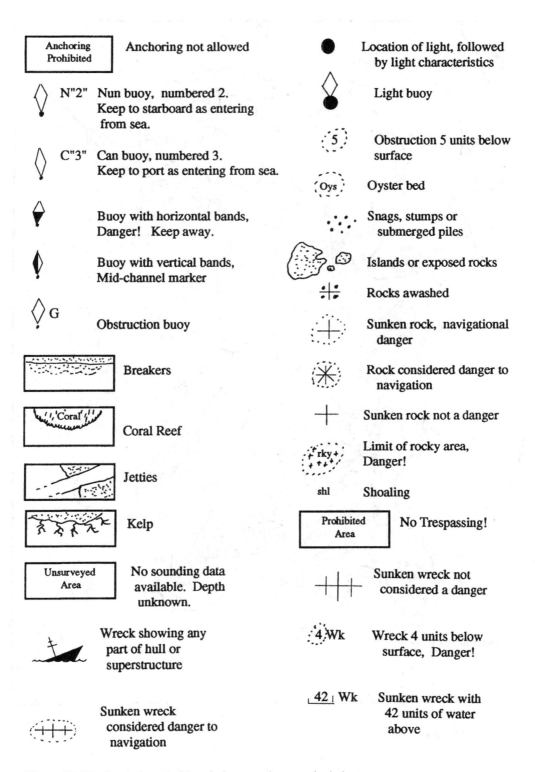

Figure 13-19.  Symbols and abbreviations used on nautical charts.

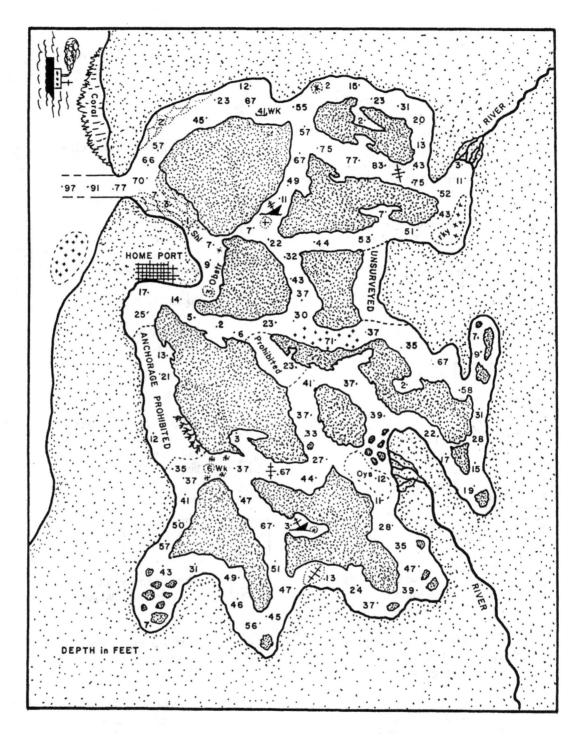

Figure 13-20. Home Port chart.

Photocopy this page onto a transparency.

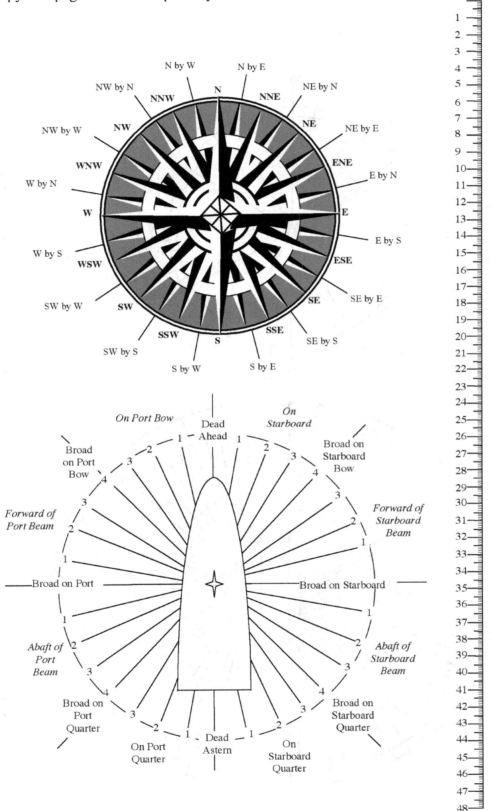

Photocopy this page onto a transparency.

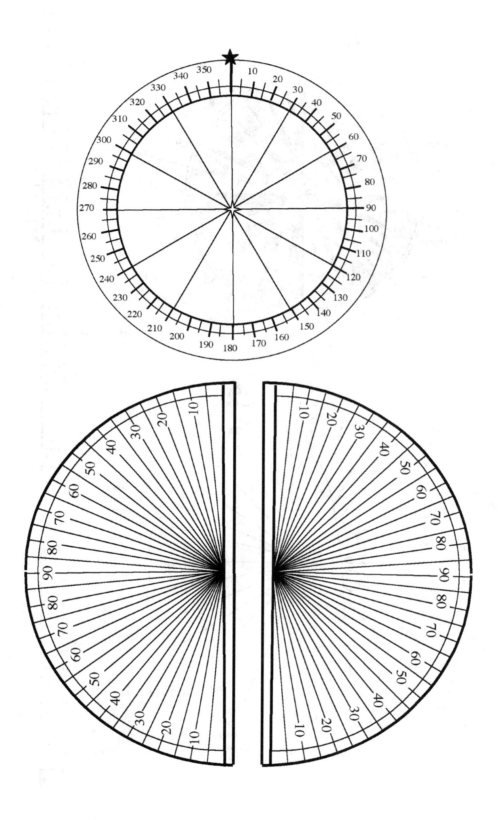